Praise for Charles H. Thornton's Book

From tall buildings to building companies, the new memoir from prominent structural engineer Charles H. Thornton shares his "just do it" philosophy and his step-by-step formula for making firms successful. He weaves tales of childhood, child-rearing and Outward Bound adventures with stories about acquiring and starting companies and designing projects around the world – storytelling from a professional.

—JAN TUCHMAN, EDITOR-IN-CHIEF, ENR

As good as "A Life of Elegant Solutions" is as a title for Charles Thornton's memoirs, the book could just as well have been called "Doing Well and Doing Good." Throughout the book, Thornton presents engineering not only as an immensely challenging and satisfying career, but also as a noble profession and a means for contributing to the betterment of the human condition. This lively portrayal of a life full of accomplishment and purpose complements Thornton's tireless work, through the ACE Mentor Program, to encourage young people to pursue careers in architecture, engineering, and the construction professions.

—BRENNAN O'DONNELL, PRESIDENT, MANHATTAN COLLEGE

The book vividly describes Thornton's life and work in structural engineering – why and how he accomplished what he did. The style of writing is terrific as I felt like he was personally talking to me the whole time. I particularly liked the descriptions of some of his projects (e.g. Petronas Towers, the collapse of the Hartford Coliseum in Connecticut, and the government complex in Oklahoma), as well as the motivation for the founding of the ACE Mentor Program. The book's well-placed harbingers give a hint as to what is coming. Most importantly, Thornton's upbeat approach to life and his zest for the pursuit of excellence and innovation in structural engineering comes across throughout the book in a highly compelling way.

—CHRISTOPHER ROJAHN, EXECUTIVE DIRECTOR,
APPLIED TECHNOLOGY COUNCIL (ATC)

It is easy to understand why Charlie Thornton has been so successful at everything he has attempted. His passion for engineering and life is abundantly clear. A fast and interesting read with several good life lessons interspersed.

—TOM GILBANE JR., CHAIRMAN AND CEO, GILBANE BUILDING COMPANY

A fast paced behind-the-scenes look at the life a man who made a difference not only to the iconic structures we've come to know in our built environment, but more importantly to the lives of thousands of mentored students who will find a career in the rewarding fields of design and construction thanks to his efforts.

—MARK A. CASSO, ESQ.,
PRESIDENT, CONSTRUCTION INDUSTRY ROUND TABLE

In this book, Charlie reflects on relationships with family, teachers, colleagues and friends who have helped shape his life and career. To those who spend any amount of time with him, it is clear that Charlie is a true and dedicated mentor. Long before he founded the ACE Mentor Program, Charlie made time to connect, to offer his insight, guidance and support to generations of engineers who have grown and are growing into leaders in the industry. This is his most lasting legacy.

—PETER DAVOREN,
PRESIDENT AND CEO, TURNER CONSTRUCTION COMPANY

Charlie Thornton has done us a favor by chronicling the story of a life well lived. It should be required reading for all of our youth today to give them a clear vision of what's possible in our great country.

—HANK HARRIS, PRESIDENT AND CEO, FMI CORPORATION

I predict it will be a "must read" for every youngster considering a career in the engineering profession.

—HARRY ARMEN, FORMER CHIEF TECHNOLOGIST,
NORTHROP GRUMMAN CORPORATION

CHARLES H. THORNTON:
A LIFE OF ELEGANT SOLUTIONS

A Memoir

By

Charles H. Thornton, PhD PE

with Amy Blades Steward

Cover design by Conner Dorbin and Deirdre Devlin Kelly.

Petronas Towers – Front cover image courtesy of Yojik/Shutterstock.com.

ACE Students – Front cover image courtesy of Michael Goodman.

Charlie with Statue of Liberty – Back cover image courtesy of Swanke, Hayden, and Connell and Dan Cornish.

All other photos are courtesy of Charles H. Thornton PhD PE.

ISBN-13: 9781495278174
ISBN-10: 1495278174
Library of Congress Control Number: 2014901310
CreateSpace Independent Publishing Platform
North Charleston, South Carolina

"Writing a book is an adventure. To begin with it is a toy,
then an amusement. Then it becomes a mistress, and then it
becomes a master, and then it becomes a tyrant and, in the last stage,
just as you are about to be reconciled to your servitude,
you kill the monster and fling him to the public."

—WINSTON S. CHURCHILL, 1942

Acknowledgements

CREATING A BUILDING DESIGN IS a positive, enjoyable and satisfying opus. Starting a family and having children is also an extremely positive experience. I've taken great pains to keep this book in a totally positive vein. While I have had obstacles and challenges to overcome throughout my life, I have intentionally left out most references to negative things in my childhood, early career, and my families in order to create an inspiring story of taking risks and accomplishing dreams.

In 2011, I attended a cocktail party for Chesapeake Chamber Music, at Penny Proserpi's house on the Choptank River in Bellevue, MD. There I met Kimberly Sedmak, a producer for AARP, who lives in Royal Oak, MD. Kim produces *Your Life Calling with Jane Pauley*, monthly segments featured on NBC's *TODAY* Show. These segments are based upon people who have spent part of their lives doing one thing and then the next part of their lives reinventing themselves and doing something else. During the cocktail party, I started telling Kim about the ACE Mentor Program, which I founded in 1995. Several months later, Kim called me and asked if I would be willing to be profiled in one of Jane Pauley's segments. Kim was intrigued by my involvement in ACE and the organization's achievements.

In December 2012, Jane Pauley, Kim, and the team produced a very effective piece about me and ACE. The piece included footage at one of Thornton Tomasetti's engineering projects, Battery Park City Winter Garden, right next to the Ground Zero World Trade Center site. They also included some footage at Manhattan College, my alma mater, and some footage at an ACE New York City fundraiser at the JP Morgan Library, designed by architect Renzo Piano on Madison Avenue in New York City.

Jane is an extremely intelligent, warm, kind, caring, and wonderful person with whom to interact. She told me about her own autobiography, entitled *Skywriting: a Life Out of the Blue.* I purchased it and couldn't put it down – it was mesmerizing. I had been planning to write a book about the real reason the World Trade Center fell, but after showing it to some close relatives, I was advised to publish it posthumously. Upon reading Jane's book, I realized it was time instead to write my memoir, *Charles H. Thornton: A Life of Elegant Solutions.* Between 2010 and 2012, I started writing outlines for the book, keeping them in a folder. After reading Jane's book, I decided the time had arrived to start my book.

Throughout my career, I had always dictated into a dictation machine, which was initially on tape and later, digital. As a result, I was always able to dictate an entire technical report on extremely complex subjects in almost no time. Mary Ann Owsley was my very able executive assistant for several years before I left New York in 2000. She was a master at reading my mind and knowing exactly what had to be done to a rough draft. I would always get in early to dictate several items. Mary Ann transcribed the tape and had drafts back to me in the afternoon, she had an uncanny ability to know what I thought the priorities were and she always did what had to be done first and left the other stuff to the end. Each evening, I would edit Mary Ann's drafts on the Metro-North train to Connecticut, where I lived. The next morning, I would bring the marked up draft into the office and the final report was generally done that next day.

Because I was comfortable with dictation, I purchased Dragon Naturally Speaking software to begin the process. I met Amy Steward, a creative writer in Easton, MD, who was the stimulus that got me revved up to start my book. Meeting Amy was fortuitous. She is a brilliant writer and has taken the time to really interest herself in my life. She has truly made the production of this book her "Elegant Solution." I would also like to thank Judy Reveal, whose expertise in editing and indexing, polished the book; and Conner Dorbin, an ACE student who helped design the cover of the book with Deirdre Devlin Kelly, a talented graphic designer.

I would be remiss if I did not acknowledge the support and encouragement that I received throughout my life from my family, including my loving wife Carolyn and my four children, Diana, Kathy, Charlie, and Becky. They have enabled me to pursue my life's dreams and to write this memoir. They also have offered support by contributing stories and photos and reviewing the book.

Finally, I have tried to include in this book all of the people who had positive influences on me and my life and apologize for any omissions that possibly may have occurred.

Table of Contents

PART THREE – RENAISSANCE/DISRUPTIVE INNOVATION

Prologue

The Ultimate Elegant Solution

THE PURPOSE OF THIS BOOK is to tell a story – my journey as a structural engineer. After working as a successful engineer for over 30 years, I want to provide a guide for high school students looking toward college and a career, for college students coping with the rigors of an engineering program, for graduate students facing a master's or PhD thesis, for a young engineer trying to figure out how to move up the company ladder, and for professionals seeking to start a business. I hope my *15-Steps to Success* will help business owners to be more successful and keep the momentum going that has made their firms successful, as well as to insure that the management of their firms continues beyond their involvement.

My mantra has been "Passion, Persistence, and Flexibility" – without these convictions in life, we won't make it. I summarize the risks I have taken throughout my career and personal life at the end of the book. I hope by telling the stories about how I managed these risks will inspire others to learn how to manage the risks in their lives.

I realized in 1991 that I wanted to do exactly what my father did, which was to "give back" and help unfortunate children get a decent education and provide them the opportunity to become hard-working citizens.

In 1991, Dr. Joseph Lestingi, Dean of Engineering at Manhattan College, my undergraduate alma mater, called me and the other Board of Advisors to Manhattan College's Engineering School and said, "Because of the end of the Cold War and the Berlin Wall coming down, defense budgets have been slashed and engineering enrollments in colleges in the North East are plunging. This plus the other events happening – the real estate crunch and the Savings and Loans crisis – has the Board of Trustees thinking of closing the Engineering School."

Well, that was all the Board of Advisors needed to hear! We mobilized and met with Dr. Lestingi and the heads of all departments of Manhattan College's Engineering School. They reported that for 80 years a system of Catholic high schools in New York City had fed white boys into Manhattan College's Engineering School. I knew this because when I went to Manhattan College from 1957 to 1961, it was all predominately white men. By 1990, the Manhattan College feeder system of white male Catholic high school graduates had dried up. The demographics of the Catholic high schools were changing and these schools were no longer graduating strictly white males.

So, the Board of Advisors at Manhattan College formed a group of companies to reach out to the traditionally underrepresented population of minority and women students to interest them in the industries of architecture, construction management, and engineering. The group consisted of myself (Thornton Tomasetti), Steve Greenfield (Parsons Brinckerhoff), Ray Monti (Port Authority of New York and New Jersey), Bob Rubin (Postner & Rubin), John Magliano (Syska and Hennessy), Ed Rytter (Chase Manhattan Bank), Robert Borton (Lehrer McGovern Bovis), and Lou Switzer (the Switzer Group). We started with the New York City public high schools and found minority men and women who might be interested in the industries of architecture, construction, and engineering.

In 1993, through John Woodman and Taip Redsovik, we found, through the New York City School Construction Authority, 350 summer interns who had jobs with companies in the industry, including my company, Thornton Tomasetti, Syska and Henessey, Morse Diesel, Turner Construction, the Port Authority of New York and New Jersey, as well as in the offices of many architects and engineers. These students worked with architects, engineers, and those in the construction industry for eight weeks each summer. As the program progressed, we realized that we needed a mentoring program to keep these students interested during the following school year between September and June. It was a natural fit for us to recruit these summer students as the first students to participate in our program during the upcoming school year.

The group evolved and in 1993, Thornton Tomasetti took the lead. Backed up by my executive assistant, Mary Anne Owsley, and engineer, David Peraza, we formed a team within our company. Soon, several other companies signed on, including an architectural firm and a mechanical/electrical/plumbing firm, as well as a construction company. In 1995, the Architecture, Construction, and Engineering (ACE) Mentor Program was formed. The program took off and between 1995 and 2000; as we expanded to Newark, New Jersey and Stamford, CT. The expression, Keep it Simple Stupid (KISS), is a model for the ACE Program. The ACE Mentor Program is elegantly simple and the ultimate elegant solution. I am dedicating the proceeds from this book to help the program grow.

What is an Elegant Solution?

The term elegant solution, in mathematics and physics, indicates a way of solving a problem that is correct and efficient, but beyond that is also pleasing to contemplate.

During the first three years of my graduate work at New York University I took many mathematics courses on the University Heights campus in the Bronx and several mathematical courses at the Courant Institute of Mathematics at the NYU Washington Square campus. It was during that time that I learned from one professor that in Napoleonic France in the early 1800s, French academies and the French government held mathematics competitions. The winner was the participant who provided the most clever, fast, efficient solution to the mathematical problem. The solution was called an "Elegant Solution." When we purchased our first sailboat, Carolyn suggested that in marketing and in computer programming, an elegant solution was also defined as a very clever efficient marketing effort or a solution to an algorithm. I liked the name and immediately we named the boat Elegant Solution.

On several occasions, while anchored out in the Nantucket or Martha's Vineyard area, sailors in other dinghies would approach our boat and ask, "Is there a mathematician on board?" Shortly thereafter, Jonmerie, a yacht builder from Finland, used the term elegant solution to describe his elegant 40-foot sloop. It was really a beautiful boat. The Tartan 37 Elegant Solution II and the Tartan 4100 Elegant Solution III that we eventually owned were both flag blue hulls and were eye catchers when coming into the harbor. Therefore, an elegant solution is not only a clever way to solve an issue but it is an aesthetically pleasing solution.

Over the years, I have read many books about San Francisco Bay and the two bridges that cross it. As mentioned in the body of this book, elegant solutions are created by collaboration between great architects and great engineers. The Golden Gate Bridge was actually designed by collaboration between Joseph Strauss and his engineer Charles Ellis, a structural engineering professor at a midwestern university, as well as William Morrow, an architect who commuted on the ferry across the San Francisco Bay from Oakland to San Francisco. Morrow observed many sunsets and sunrises, which led to his suggestion for the coloration of the bridge and the Art Deco detailing which enhances the bridge's aesthetic appeal. The beautiful Golden Gate Bridge is a great example of an elegant solution, while the rather pedestrian Oakland Bay Bridge is not so elegant.

These are very simple examples of elegant solutions – yachts and suspension bridges. In my professional career, I have always strived to make all of my projects elegant solutions. I hope this book will inspire others to pursue their life's passion and seek their own elegant solutions to the challenges life presents – and to have fun along the way.

Introduction

"The future is not a result of choices among alternative paths offered by the present, but a place that is created – created first in the mind and will, created next in activity. The future is not someplace we are going to, but one we are creating. The paths are not to be found, but made, and the activity of making them changes both the maker and the destination."

— JOHN H. SCHAAR, *LEGITIMACY IN THE MODERN STATE,* TRANSACTION PUBLISHERS, 1981.

IN THE SUMMER OF 1991, Jon Pickard, an architect in the architect Cesar Pelli's office in New Haven, CT, called me and asked me if my firm, Thornton Tomasetti, would join their team to pursue a design competition for a tall building in Kuala Lumpur, Malaysia. We backed up the Pelli proposal with some compelling structural diagrams to help them win the project. We won the competition and our first trip to Malaysia occurred in November of 1991.

We proceeded with multiple structural and architectural concepts, we priced them, we scheduled them, and we educated our client as to why the scheme that we were proposing was the right scheme. I knew that the steel industry in Taiwan and Japan, as well as Korea, would be pushing an all steel building. But this was the wrong solution for this project. The right solution was a high strength concrete core and circular perimeter concrete tube and a simple steel floor system, similar to the Miglin Beitler structure in Chicago that we had already designed, but was not built. We educated everybody at Petronas, the Malaysian National Oil Company, and Lehrer McGovern Bovis, the construction managers, that the scheme we proposed was in fact the right one and we proceeded with the final design.

The international design competition for the Petronas Towers, called Kuala Lumpur City Center (KLCC), did not specifically call for twin towers or the world's tallest buildings. It was a master plan done by Kalegis-Vales, a California company that gave some rough guidelines. The program called for a large mixed-use office, retail, parking, and hotel uses. As we commenced our work after our initial trip to Kuala Lumpur in late 1991, the twin tower concept evolved. Working closely with Cesar Pelli and Jon Pickard, Bob DeScenza, Len Joseph, Paul Lew and Udom Hungespruke of Thornton Tomasetti developed the twin tower concept – each tower containing 88 stories connected together by a bridge at about the 42nd floor. The original intent was to use the bridge to reduce motion and damp cross wind vortex shedding vibrations under wind loads. The twin towers would sit on a podium which contained parking for over 6000 cars and several million square feet of retail and theaters. The towers would be 2,200,000 square feet each for a total of 4,400,000 square feet, but the entire project encompassed 12,000,000 square feet.

After we won the competition, we actually got a chance to see the other seven competing designs during the first trip to Kuala Lumpur. Most of the competitors took projects they had designed in other places that had been either built or not built, and they stuck them together into a submission. Only Pelli and Pickard's designs included textures, shapes and geometry indigenous to Malaysia. The Pelli design was far superior to anything else.

Putting things in perspective, very few structural engineers ever get the opportunity to design the world's tallest building. The Empire State Building, designed by pioneering structural engineer Homer Gage Balcom, was built in 1933 using a structural steel frame. It remained the world's tallest building for 40 years until 1973, when Dr. Fazlur Kahn of

Skidmore Owings and Merrill in Chicago got the next crack at building the world's tallest building – the Sears Tower. It was also a structural steel system. About the same time in 1974, John Skilling, a Seattle-based structural engineer, became the structural engineer on the World Trade Center in New York City – also a structural steel system. It held the record as the world's tallest building until Thornton Tomasetti topped out Petronas Towers in 1996 after a period of 22 years, using our composite concrete and steel approach.

After we won the job in Kuala Lumpur, we had a chance to take the submitted concept to a more complete design. Then came the big day in 1992 when the team, including Cesar Pelli, Jon Pickard, Norman Kurtz of Flack and Kurtz, mechanical/electrical engineers, Gene McGovern of Lehrer McGovern Bovis, Fred Clark and Larry Ng of Pelli and I, had the opportunity to join Cesar in presenting multiple schemes to Malaysia's Prime Minister, Dr. Mahathir Mohamad; the president of Petronas (the Malaysian National Oil Company); the Mayor of Kuala Lumpur City Centre, Azizn Zainul Abidin; and other dignitaries.

Cesar Pelli is a master of presenting beautiful designs in a compelling manner. I had seen him in action on the 72-story Norwest Tower in Minneapolis, MN, and the Mayo Clinic in Rochester, MN. He is a master of understatement and calmness! Cesar started with the first of multiple models and proceeded with the balance of the models, which progressively got taller and taller. Each time he placed one on the table, the Prime Minister smiled and Cesar smiled back very demurely. When Cesar lifted the last and tallest model, the Prime Minister beamed and said, "Would that be the world's tallest building?" Cesar smiled and said "yes." The Prime Minister then said, "Can you (we) do it?" Cesar Pelli then turned to me and said, "Charlie, can we do it?" Without hesitating, I said "Absolutely!"

Although this was not rehearsed, in my mind I had rehearsed this scenario many times over the years. Taking a risk was just in my nature from the beginning. These risks led to the accomplishments in my life.

PART ONE – ORIGINS

1
The Formative Years

"At times I count them over and over, every one apart – my rosary of days lived long ago in that land where, for a few short years, time had stood still – days that merge and blend and weave themselves into a colorful tapestry of months and years as I journey back down the half-forgotten road to yesterday – a journey that leads to a golden land of dreams and memories, a sanctuary filled with hopes and dreams and aspirations that will never die."

— JAMES K. PEIRSOL FROM "MENDOCINO: A PAINTED PICTORIAL" BY KEVIN MILLIGAN, COASTSIDE GRAPHICS, 2002.

Childhood in Clason Point, Bronx, New York

MY FIRST HOME WAS 252 Leland Avenue in the Clason Point section of the Bronx in New York City. I lived there for 21 years. Located on the East River, it was a turn of the century converted waterfront bungalow that had been jacked up and brick-veneered by my father, Charles, and all of his father's cousins, Eddy, Paul, Artie, Leslie, Hugh Sr., and Bill Thornton. Clason Point was a neighborhood of blue-collar workers right on the flight path to LaGuardia Airport. My maternal grandfather, Frederick

Heinemann, always referred to Clason Point as a backwater attachment to New York City. He called all us boys sticks-in-the-mud.

Fred Heinemann's parents came from Germany in the 1800s. On my paternal grandfather's side, three brothers, all masons and bricklayers, came to the United States from Northern Ireland in 1840. My father was born in 1905. His father died less than six months later, and his mother "disappeared." He was raised by Aunt Marnie and Aunt Mamie, two "lace-curtain" Irish spinsters who were Thorntons and supported by a branch of the family called the Reillys, wealthy people who resided in Mount Vernon, Westchester County. At age 29, my father was approached by his aunts and asked if he wanted to meet his mother, whom he thought did not exist. He chose not to meet her.

My father had a tough childhood, but that didn't bother him. He was a tough, tough guy – a runner and boxer in the National Guard. As a young man, he worked his way from electrician to bricklayer. In the early 40's he came in from the cold and became a supervisor on the night crews on the Queens/Midtown tunnel. While on the job, he had a serious accident and was scalded by hot asphalt tar. I remember him coming home looking like The Mummy. He then took the civil service tests and became a New York City construction inspector and between that time and 1969, my father became Chief Inspector of the New York City Building Department in the Bronx.

Between 1954 and 1961, I watched my father's actions in our neighborhood, getting every high school drop-out into the construction apprentice programs for carpenters, bricklayers and laborers. There was quite a bit of corruption in those days and my father's bribes were jobs for kids. When I was a child, we really didn't have much. My father generally made us work every weekend moving piles of bricks and yards of sand and gravel. It amounted to a luncheon allowance for the work I did. My father bought a vacant lot next door to us and he, along with relatives, and my brother Bill and I, built a two-car garage on it. We worked every weekend on some project – whether it was adding a front porch or a bedroom addition to the back of the house. My friends, John Kovac and Charles, Hughie, Jack, Terry, and Danny McKinney, would come calling for me to go do something and I would say that I had a job to do, but if they would help me, I would be available sooner. It was like Tom Sawyer painting the fence – the beginning of my leadership training.

At the turn of the century, Clason Point included Harding Park, a bungalow colony surrounded by open fields. Harding Park was once referred to as Bronx Brigadoon in a prominent architectural magazine. Clason Point was bound on the north by Bruckner Boulevard, on the east by Eastchester Creek and Pugsley's Creek, on the west by the Bronx River, and on the south by the East River, the extension of Long Island Sound.

I grew up hunting rabbits and pheasants with BB guns and bows and arrows and fishing for eels, flounder, and other types of fish. We also kayaked and went power boating to Long Island's North Shore. From the front porch of our home in Clason Point, we could view the New York skyline through the belching black smokestacks and piles of coke belonging to the nearby Con Edison Coke plant. The plant ultimately gave way to the Hunts Point Cooperative Market, the largest food distribution center in the world, which is still in operation today. The odors from the Bronx River to the west were ground coffee aromas coming from Café Bustelo Coffee's roasting plant. Thank God the prevailing winds were from the west, because the odor emanating from Pugsley Creek at low tide was that of rotten–eggs or sulfur and it stunk! The Bronx was a tough place and there was an adventure around every corner. The landfill, which ultimately became Soundview Park, was covered with topsoil and planted with grasses; we would play hide and seek in the Phragmites in the fog.

Clason Point Park was known for its amusements and entertainment, as well as the world's largest saltwater outdoor swimming pool, known as The Inkwell. Between 1900 and 1970, the people who owned the bungalows did not own the land on which they stood. They were essentially squatters, paying a land lease to whoever owned the property. The owners varied between a hospital and a developer. During World War II, I remember the flying boats of Pan Am World Airways landing in the water of the East River or at the Marine air terminal, which preceded LaGuardia Airport.

In 1949, urban planner, Robert Moses, gained a power of new immensity, called Title One of the Housing Act of 1949, which extended the power of eminent domain. One of Moses' targets for redevelopment was Harding Park, named after President Harding. Harding Park survived the attempt of Robert Moses to tear down what he called the Soundview Slums but because it was nearby, I lived for over five years in Clason Point with the threat of being evicted by Title One. Author Robert Caro, in his book *The Power Broker*, explains that urban expert Charles Adams said that under the new redevelopment laws, Macy's could condemn Gimbels, if Robert Moses gave the word. To this day, my fear of the government taking our house has always been real. I loved Clason Point because it overcame that threat and survived.

In the 40s and 50s, families in Clason Point fell into two family types – the haves (those who had a father in residence) and the have-nots (those who had no father at home or alive). There was only one other family, my immediate neighbor, who was a have – the Kovacs. John Kovac had a mother and a father, a member of Electrical Union Local 3, in residence and they both worked. John graduated from high school and went into the Local 3 – the toughest union to break into unless you were both white and the son of a member. All the rest of the families in Clason Point – the McKinneys and the Haugheys – were have-nots – in almost all cases the father was either dead or not capable of earning a living. My father tried to convert these families from boys to men; earning a living by helping them find jobs as bricklayers or carpenters in the New York construction industry. As my role model, my father's example encouraged me many years later to start the ACE Mentoring Program in the NYC public schools to attract high school students to careers in architecture, construction, and engineering, focusing on minority men and women.

My family included my mother Evelyn (b. 1916), father Charles, older brother William, (b. 1938), and younger brother Robert (b. 1945). I was born in 1940. Although my name was also Charles, my mother called me "Buddy," because I was my big brother Bill's "Little Buddy" and they didn't want another Charles, by name, in the house. My father worked for New York City as an electrician, a bricklayer, and finally as a construction inspector for New York City's Department of Buildings. He went to work early and came home early. I relished his arrival back home each day. Usually I would be sitting alone or with my brothers on the kitchen floor, playing with trucks or toys, and he would sweep into the kitchen and lift my mother off her feet, kiss and grab her all over. My mother would squeal with laughter and say, "Charlie, the boys are watching." It was true love. I don't think I ever saw my parents have a cross word between them. They were the perfect parents – they paid attention to their children and they loved each other.

When I was about ten-years-old, I was at a family party at a cousin's house on Leland Avenue and I walked up to a woman at the party, smacked her on the butt, and said, "Just like my mother's." Of course, I was the center of attention – the precocious little bastard. I truly believe that birth order has something to do with personality. When you are born a year and a half after your big brother, your behavioral patterns lead you to be precocious – otherwise, no one is going to notice you. Despite that, I had a very happy childhood. Even today, my father's cousin, Hugh Thornton, said that I haven't stopped smiling since I was a child.

As I grew older, by age 10, the children of Clason Point tended to hang out in cliques. The boys had their groups and the girls had theirs. It became an unofficial rule that if you were with a friend and someone came along and tried to horn in and you left with that person, you were called 'flat leaver.' In other words – you were disloyal. The girls were generally much worse than the boys about doing this.

All my friends and I had bicycles, whose parts were salvaged from hand-me-downs and wrecks, put together by my father. Not one of them had more than one gear! Some of us had balloon tires and others had skinny tires. We thought nothing of jumping on our bikes and riding the 29 miles from the south end of the Bronx River Parkway north to White Plains and back, which took us three hours each way. We also rode our bikes to Orchard Beach, which was in the northeast section of the Bronx. This was probably about a 10-mile trip each way. We all had very strong legs. Hunter's Island was adjacent to Orchard Beach, so we would often carry small backpacks with Sterno stoves and cans of soup, and camp on the island.

I was always kayaking, boating, rafting, and doing things related to the water. Eileen Simmons was a girl about three or four years older than me who lived on Soundview Avenue near the local hardware store. I don't remember the circumstances, but Eileen decided to take me into the Pugsley Creek swamp. Many of these marshy areas surrounding Clason Point had mosquito ditches to drain the swamp and eliminate the mosquitoes. No one knew where we were that day. While jumping across mosquito ditches, I stepped down and sank, pulling my foot out to find I had lost my shoe. So I hobbled back to the dead-end street that led to the Pugsley Creek swamp, and to my surprise there were about 15 people there, including my mother and father. So, Shoeless Joe Jackson returned without his shoe. This did not make my parents very happy and they did not like the idea that at age six I was lost in the swamp with a girl three to four years older. Just another adventure!

Life on the Water

When I was eight-years-old and my brother Bill was ten-years-old, my father bought Bill and me a 10-foot long sailing kayak. It was a wood frame covered in canvas with a lateen sail rig, a triangular sail mounted at an angle on the mast. We didn't use the sail much, but we did use the kayak. My father encouraged us to explore rivers around Clason Point. As long as my father seemed to know what we were doing, my mother said it was all right with her. Some of my favorite memories are of kayaking on the East River. We kayaked all around the East River in the vicinity of Rikers Island and the Whitestone Bridge. Many times we would be kayaking under the Whitestone Bridge, with no lifejackets on, dodging freighters and tankers. During my mother Evelyn's last years, when she lived near us in Easton, MD, I used to tease her that she was a bad mother because she allowed us, at such a young age, to kayak in such dangerous waters. She vigorously denied any knowledge of our escapades. I guess this is how I learned to take risks and not have them get in my way.

My grandfather, Fred Heinemann, was a cabinet maker and a master carpenter who had very sharp tools. My father's tools were all pretty dull as he left them everywhere and didn't take care of them. My grandfather helped me, my brother Bill, John Kovac, Jack McKinney and other kids in the neighborhood to build boats. We actually built five kayaks in our garage. One of the boats we built was called "Leaky Lou." It was a clear pine shipping crate that we found in the city dump along Bruckner Boulevard. We were always in the city dump picking up four-foot by eight-foot sheets of 3/8-inch exterior plywood from construction debris, two-by-fours, four-by-fours, and copper wire.

During this time, we also sold newspapers for a penny a pound and copper for about a dollar a pound. We made a lot of money scrounging in the city dump and bringing it back and selling it to the local junkie. A junkie meant something different in those days – it was a guy on a horse and wagon who came by and shouted, "I buy papers, I buy junk." We even had an ice man with a horse and wagon who delivered ice, as a lot of people didn't have refrigerators yet. This scavenging helped fund our boat projects.

The next boat John Kovac and I built was an eight-foot hydroplane. It had plywood sides and bottom on a wood frame, with a canvas deck painted with two lightning bolts. We christened the boat "Fantasy II." John Kovac's father bought a Scott Atwater seven and half horsepower engine with a Bail-o-matic, and the boat skimmed across the water at about 25 miles an hour. The problem with this boat was that it leaked. We really weren't very good at caulking the hull. The next boat we built was a 14-foot runabout. One of the best trips we did in this boat was going wide open through Hell Gate, a narrow tidal strait in the East River. We cracked a rib of the boat on that trip from bouncing in the rough waves in this section of the river.

In 1955, my father was able to buy a 15-foot Lyman powerboat – a really classy boat built in Sandusky, OH made of 3/8-inch lapstrake and brass rivets. We used this boat to explore Rikers Island, the Whitestone Bridge, Ferry Point Park, the Bronx River, Westchester Creek, Pugsley Creek, and all the way out to the Throgs Neck Bridge and City Island. This is the boat we took out into hurricanes during the mid to late 50s. When every hurricane hit, many of the boats that were anchored at Willets Point would break loose and end up on Rikers Island. One or two times, we would go out in the eye of the storm when the sun came out. We thought the hurricane was over and then it would pass by us again. We got into trouble and would go up on the beach of Rikers Island and pull the boats off that had broken loose before the police and their huge German shepherds could get us. We would high tail it back to my house. Most of the boats came from Captain Bill's and he would find out that the Thornton crowd over in Clason Point – "the Pirates" – had possession of several of his boats. He would come with a truck and pick up the boats. We netted one boat per storm.

Every weekend, the Kovacs, McKinneys, and my sister-in-law Joan Thornton's father Nick, Ernie Howe, and the Yacht Club members would go to the North Shore of Long Island, called The Gold Coast, to hang out with fellow members of the Bronx Corinthian Yacht Club. We would all head out to Manhasset Bay, Plum Point, and Half-Moon Beach on Saturdays and Sundays and hang out below high water, because the elites could not chase us if we stayed below high water. We also started going to City Island all the time with some of our friends and girlfriends. City Island was the yachting center of the East Coast and the home of many boat yards, yacht building facilities, and sail makers. In fact, most of the America's Cup boats of the 50s and 60s were constructed in yacht building facilities on City Island.

We also traveled in the other direction, down through Hell Gate, down the East River, and out through the Narrows around to the Sheepshead Bay, where we would go into a marina for soft ice cream. This was actually lunacy, because if we went down in the evening heading west there was brightness in the sky, but coming back heading east it was pitch black. We used to go through Hell Gate and the East River wide open at 25 to 30 miles an hour and couldn't see anything. I didn't realize until years later that this is something we shouldn't have done.

One night we lost John Kovac. He and I were both out in our boats with our girlfriends. As we went under the Hell Gate Bridge, he went on one side of a tanker in his boat and I went on the other side of the tanker in my boat. When I came out on the northern end, he wasn't there. I then went home and woke up everybody and told them he was missing. We launched a search, but there was no sign of John and his date. Little did we know, John had run out of gas and did not want to be discovered. He was anchored off the northern end of Roosevelt Island. I'm not sure what he was doing, but his boat didn't have its lights on. John's father noticed a reflection off his chrome hardware and was fit to be tied.

One of the best boat trips involved my father, Bill and me in the Lyman. We went all the way out past Port Jefferson to Rocky Point, probably 50 miles each way. We actually saw a large dolphin or porpoise right next to the boat as we came into the beach. Unexpectedly seeing one up close really scared us. Sleeping on the beach was less than glamorous. We got eaten alive by sand fleas that night and we returned home the next day.

As my life progressed, I started college in 1957. I got married in 1961 and we left the Lyman in the care of my younger brother, Robert, who didn't take very good care of it. It warped over the winter and we junked it. This ended my boating career until I attended a boat show where I bought two 17-foot canoes – a Grumman sailing canoe and a Blue Hole ABS white-water canoe. This is when I began sharing adventures with my own family.

The Boy Scouts

From 1948 to 1951, my father decided to become the scoutmaster of Boy Scout Troop 261, which met in a Protestant church in my neighborhood. It was a tough neighborhood and we actually were accosted by a lot of the kids who were unable to be Boy Scouts. We always took a back alley and snuck away from home to go to the church, rather than invoke the wrath of some of the bullies in the neighborhood. Both my brother and I moved from Tenderfoot to First Class in the Boy Scouts. I ended up achieving the Life rank. In the Boy Scouts, we did a lot of hiking. The worst experience of my life was winter camping at Alpine, New Jersey, freezing our you-know-whats off. That experience was enough for me to decide that winter camping was not for me. In 1951, when I was only 11, my father took me and the older boys on a 40-mile hike along the Delaware River near Callicoon, New York. This was only for Explorer Scouts, ages 12 and older. I was only 11! We left at noon one day and camped out two nights and arrived at noon on the third day. The weather while we were hiking was hot and humid, and most of the springs had gone dry. With a 40-pound pack and little freshwater, it was not a very pleasant experience. It turned me off completely to backpacking.

At one point during the hiking trip, we climbed up a sheer trail leading to the cliffs overlooking the Delaware River. One of the scouts, Charles Christian, had 25 comic books in his pack, which must have been heavy. Half way up the steep trail, he laid down on the trail and said, "Please leave me here to die." After the drama, my father said the equivalent of "Get off your ass and get moving." Charles complied! My father always believed in finishing what you start, even before Nike came up with their slogan, "Just Do It." On the second evening, we found a stand of pine trees with a thick bed of pine needles. It was adjacent to a grocery store that had white birch beer. I can still taste it. We all slept like logs. This trip probably led me to canoe camping, where Mother Nature carries your equipment down the river. Once I started sailing in boats with large cabins and comfortable bunks, I called it luxury camping.

Lake Peekskill

In the summer months, my family spent time at Lake Peekskill in Oregon Corners, Putnam Valley, NY. In 1945, my father bought 10 lots in Lake Peekskill and that summer we pitched an eight-person tent and a four-person tent and spent four to six weeks exploring the great outdoors. Ultimately, my father erected an eight-foot by eight-foot construction shanty in a remote spot and we spent every summer with no heat, no air-conditioning, no water, no electricity and no outhouse or toilet. Every day, we walked to a spigot and filled up one-gallon bottles with water and cooked on a kerosene stove. We dug a hole and filled it with blocks of ice to protect our food. Going to the toilet involved taking a foldable World War II foxhole shovel and a roll of toilet paper into the woods, digging a hole, and 'doing your thing.' For boys, this was not a big deal, but for girls it was another story. Once a week we went down to Peekskill Hollow Brook with our towels, shampoo, and soap and took a bath. I couldn't name a woman today that would do that for 10 years as my mother did. She never even complained. There was no TV or radio. Every day we hiked and climbed the hills and every night, we lit a fire and told stories around the campfire, going to bed early.

We shot BB guns during our young years at Lake Peekskill, but by 1956, Bill and I wanted to shoot real guns. When I was fourteen-years-old,

I got my first .22 rifle. We created a shooting range in the basement of the Clason Point house where we locked the basement door so my mother wouldn't step into the shooting range. My father told us only to use shorts but we would use the long rifles. After this, we told our father that we needed more land. In 1956, my father went up to the Catskills near Cairo, NY, and bought 80 acres along Catskill Creek. The property had a hundred-year-old house with no amenities – no heat, no water, no windows, no electricity, and no toilet. The next five years, with the help of my mother's father, who was a master carpenter, we rebuilt the house. My grandfather built all the double-hung sash windows right on site and my father, Bill and I built two chimneys and fireplaces, drilled a well, and put in a kitchen. We put a brick-veneer on the outside of the entire house. We swam in Catskill Creek and hunted squirrels and rabbits, thoroughly enjoying our time there through the 60s. These outdoor experiences with my family and friends helped me discover my own problem-solving skills and an inherent passion for engineering.

2
The Influence of Family on Education

School Years

MY FATHER AND HIS BROTHER, my Uncle Bill, both valued education; they were extremely well read and could quote from the classics. I spent my grammar school years at PS 107 from first grade through fourth grade. Due to enrollment fluctuations, the school system transferred us to PS 69 for fifth and sixth and then back to PS 107 for the seventh and eighth grades. These were great years – there was a high quality educational experience and the schools were racially diverse. Sometime in 1953 I applied to Cardinal Hayes High School in the Bronx – some of my friends were going there and that's where I wanted to go. A year earlier my brother Bill had entered Iona Prep and was a freshman. Iona Prep was $30 a month or $300 a year and Cardinal Hayes was $10 a month or $100 a year. My parents didn't have either, so my Uncle Bill and his only sister, my Aunt Eleanor, quietly supported Bill and me to go to Iona Prep which changed my life. Uncle Bill also provided my parents with a car every time he bought a new one.

Aunt Eleanor worked for the New York Board of Education as a teacher and Uncle Bill was a captain in the New York City Fire Department.

Aunt Eleanor knew Brother Heerin, headmaster at Iona Prep, and insisted that I join my brother Bill there. I took the entrance exam and passed, even though the essay almost held me up. The essay question was to write about "How little people also count," so I wrote about how children could sit by a pond and count the ducks. It hit me when I walked out of the exam that they were really asking a different question. I guess with my literal mind, I was destined to be an engineer.

Iona Prep was on the campus of Iona College on North Avenue in New Rochelle, New York, and I was still living in the East Bronx in Clason Point. It took approximately two hours each way to get to school. The Soundview Avenue bus connecting to either the Bruckner Boulevard bus or the Pelham Bay Line IRT got me to Pelham Bay Station. Mr. Joseph Lamas, the high school football coach and history teacher, then picked up the group of students and drove from Pelham Bay Station to take them to Iona Prep.

On the first day of school in September 1954, I arrived with my khakis, sport jacket, white shirt and tie and observed the affluent Catholic students from all over Westchester and Fairfield counties arrive in brand new Chevrolet and Ford convertibles. This was quite a difference from the East Bronx. I would say that all the wealthy Italian and Irish contractors and businessmen sent their sons to Iona Prep. It was an interesting mix of ethnic groups who were mostly Catholic. It was ironic that I was there because my father never went near any church and my mother was Lutheran. Aunt Eleanor had the Catholic upbringing and was of the opinion that she knew what was right for us.

Iona Prep was a four-year all male day school, with no residence halls, so we were all "day hops." There were approximately 400 students, 100 per class, with outstanding teachers, strict discipline, and a wonderful education system. During the four years I excelled in algebra, geometry, trigonometry, solid geometry, and many other subjects. My highest grades were in English and languages, including Latin and French, and I was in the French Honor Society. Of course, I then became an engineer.

The formal training and quality of the teachers, along with the intolerance of the faculty to take any crap from anybody, made it a wonderful school. When anybody got whacked, the entire class cheered. There was one teacher, a Korean War veteran, who was one tough Marine. So a student stood up as the teacher was criticizing the class and said "That is the pot calling the kettle black." The teacher then approached the student, saying, "What did you say?" The student then repeated himself saying, "That is the pot calling the kettle black." The teacher then struck the student, sending him down to the floor. We cheered. When he stood up again, the teacher asked him again, "What did you say?" The student responded again, "That is the pot calling kettle black," and the teacher whacked him again. When the student fell to the floor the third time and was asked, "What did you say?" he replied, "Nothing sir." It was a great lesson I learned that day – knowing when to shut up.

After eight years in a co-ed public school system, where there was an anti-intellectualism attitude on the part of the male students, it was finally cool to be smart. Since those days I have read widely about the state of American high school education and the entire anti-intellectualism that is rampant within certain minority groups. In fact, at PS 107, I was always the last boy standing in a spelling bee and on the other side of the room were generally three or four girls. My crowd of friends used to call me various and sundry names like sissy and fag. Finally being in single-sex education at Iona Prep was excellent for a rather insecure boy between ages 13 and 16. There was no need to impress girls so I could just excel academically and socially. The commute time, however, limited my involvement in extra-curricular activities. The one thing listed in my yearbook for extra-curricular activities was driver's education.

After my brother Bill got his driver's license and was in his first year at Manhattan College, he drove me to school. I didn't turn 17 until April that year, so Bill dropped me off at Gun Hill Road and Bronx River Parkway where Roger Muller's father would pick us up and drop us off at Bronx River Parkway and Cross County Parkway on his way to work in White

Plains. He thought Bill Haley's song, "Shake, Rattle, and Roll" was saying "Shake Marilyn Monroe." Roger and I then hitchhiked to Iona Prep. We had on our maroon and gold Iona Prep jackets, so we were always picked up by Iona College students. On April 11, 1957, I got my driver's license and completed the year driving to and from Iona Prep.

In the early years when my brother and I attended Iona Prep, we would take different buses and trains and generally arrive at Pelham Bay Station within 15 minutes of each other. We did this because I was a pain-in-the-ass little brother and I used to heckle him on the bus as he was trying to "make friends with the girls." Rather than put up with my heckling, he left 15 minutes earlier than I did. The students at Pelham Bay Station couldn't understand why we didn't come together.

My group of friends who rode with Mr. Lamas was quite interesting. My classmates were David Rosenfeld, grandson of Morris Rosenfeld – the great America's Cup photographer – and Charlie Ulmer, son of the sail-maker from City Island and founder of Ulmer Kolius Sailmakers. David ended up going to Manhattan College. Charlie went to the Naval Academy in Annapolis and became an Olympic sailor. Several other Iona Prep graduates ended up at Manhattan College with me, including Phil Hurzeler. The point of this is that all graduates of Iona Prep went to college. That was the scheme set forth by my father, aunt, and uncle. No one else from my neighborhood in East Bronx went to college and my parents were determined that all of us would go.

My father was kicked out of grammar school in the seventh grade and at age 14 went to work as an electrician and ultimately a bricklayer. He did some high school equivalency type study. Sometime in the mid-1920s, he enrolled in the General Society of Mechanics and Tradesmen's school on 44th Street called the Mechanics Institute. Founded in 1789, it is one the oldest organizations in the United States and is modeled on the English guild system. It is a free school and offers night classes. In 1927, after three years, my father graduated with a certificate. The coursework included architectural drafting, structural detailing, blueprint reading, construction superintendent training, heating, ventilation and air-conditioning design, and plumbing design. I believe my father's most proud accomplishment was that he joined the General Society of Mechanics and Tradesmen and ultimately became president of the Society in 1968. During my high school years, he invited my brothers and me to join him and my mother at several of their meetings.

When I was about 14 and my brother Bill was 15, my father was taking college courses at Columbia University in geology. He was about fifty-years-old then and the 18 to 20-year-old students called him Pop! I believe he took over 100 credits – geology was his real hobby.

In about 1954, my father approached Bill and me and said, "Look guys, I have to do a colored map of the geology of Pennsylvania. If you do it for me, I will take you on the college's senior trip in April to study the geology of Pennsylvania." We got him an A on his map! The trip was a four-day and three-night bus trip with 20 Columbia University students – all men. I believe that Columbia University was also not co-ed at that point. On the trip, we went to Sunbury, Huntington, and other towns, staying in hotels and eating with the faculty and students. One of the students had a guitar and would play as we drove through the Pennsylvania countryside. We all had our geology tools and we studied shale and all the geological formations. It wasn't until later that it dawned on me that this was part of Dad's plan to insure that we went to college.

My brother Robert, who is five and a half years younger than me, followed Bill and me around everywhere. He was somewhat of a pain in the neck. Robert hung out with the younger McKinney's and another whole gang of kids his age that I didn't really know. Unlike Bill and I, he did not attend Iona Prep, but instead attended St. Helena's High School in the East Bronx. Bill and I were in college when Robert was in high school. Somehow my parents bribed Robert to go to college in California. They bought him a two-year old Ford Fairlane convertible and he drove to California to live with Aunt Eleanor, her housekeeper Laura, and Fred Robinson, Eleanor's adoptive father, in a nice house in North Hollywood. Robert hated it there because Laura put a padlock on the refrigerator and as a growing boy, he couldn't even get a snack at night. He enrolled in Los Angeles Valley College and let his required number of credits drop so he could be drafted and leave the house. He was drafted in 1966 into the Army and served one year in Vietnam repairing radios for the 11th Armored Calvary. When he returned to the US, my father helped him get a job with New York Telephone, where he worked for 32 years.

So early on, I recognized the value of a really quality education. All of my four children, Diana, Katherine, Charlie and Rebecca, have gone to outstanding schools, such as Gettysburg College, Babson College, University of Hartford, Washington College and George Washington University.

Dating

Mostly in our high school years, we were interested in cars – repairing cars, driving cars, racing cars, and cycling. My brother, Bill, especially liked cars and wanted to know how they worked. I just wanted to turn the car key and have it work. My father's cars, however, broke down every place we went. He got his cars from family and friends and only put oil in them – nothing else. Two favorite stories about my father's cars were that he bought my brother Bill new tires after noticing his car needed a new set of tires, but stuffed his own bald tires with hay to drive to Florida one year. Another story involved my father driving a car with no reverse gear for an entire year. This drove our grandfather, Fred Heinemann, wild. Our grandfather was German and quite a perfectionist when it came to his cars and his tools.

When we weren't working on cars, we just hung out as a group. None of us really dated any girls. My cousin Janice lived in Park Chester, a lily-white large apartment complex owned by Metropolitan Life Insurance. It was a city within a city. It had a Macy's, a movie theater, shopping, parks, and gardens – and a plentiful supply of very attractive young women.

At 16 years of age, Janice was five feet, ten inches tall with long blonde hair and was dating a 23-year-old man, much to the chagrin of her parents. She spent time with us during the holidays and summers at Lake Peekskill and was like one of the boys – one year winning a fishing contest at Orchard Beach and another year learning how to shoot our guns. Because we didn't have too many attractive young women in my neighborhood, Janice set up introductory dates with the girls of Park Chester. She concluded that we really were, as my grandfather used to call us, "stick in the muds." So she would bring girls down to Clason Point and two of us boys would hold one of our friends down while the girl showed the boy how to kiss.

In the summer of 1957 after I graduated from Iona Prep, I met Patricia Podaski at the Bronx Corinthian Yacht Club, part of White's Boatyard. Patricia was a high school senior from Walton High School, an all-girls high school in the Northwest Bronx. My brother's wife, Joan Kossoff, attended junior high school with Patricia. Every Saturday and Sunday, a group of 17 guys and 17 girls would head out for a ride in our boats. The water was clean and the swimming and fishing were great. We had beach parties sometimes during the day and sometimes during the night, with fires on the beach until we got chased. Occasionally during this period, Patricia had to go and get some treatments for an illness. She was never specific about it and we never discussed it. This turned out to be radiation treatments for her lymph nodes. The treatments worked, but she had a recurrence about every five years. She refused to tell anybody about it.

Summer Jobs

In 1954 when I was 14, my father took two weeks off and got us a job helping to construct an A & P store in the East Bronx to show me how to become a bricklayer. At 14, this was child labor. I remember working in the hot summer sun. It was 96 degrees and I looked around for my father. He was nowhere to be seen. I looked across the street and he was sacked out on an earthen berm under a tree. When I quit for the day, I walked over and asked him what was going on? He said, "You have to work, I don't."

During the summer of 1955, I worked as a carpenter's apprentice renovating the Union Square Bank building at 15th Street and Park Avenue on the east side of Union Square. I worked the whole summer. One of my major jobs was lying on my back and chipping out an entire plaster ceiling. All the plaster and dust fell on my face and in my mouth. I felt like Michelangelo. I also remember carrying over one hundred 96-pound bags of cement from a truck into the building. My father's theory about hard labor and it making me stay in school was beginning to take hold. This was my first encounter with New York City construction unions. My father got me a job with the superintendent, but everybody on the job knew that I was not an apprentice and they would harass me and say, "Hey kid, we know you are not a member of the union" and "Where is your union card?" I would say, "Oh, well, I will get one tomorrow." even though I didn't have one. I worked very hard. When I rode the Lexington Avenue subway from Pelham Bay Park down 14th Street, everybody moved away from me when I raised my arm to hold onto the strap. Later in life, I visited a construction site in Sydney, Australia and they had lockers and showers and everybody left the job site, showered and clean-shaven.

In the summer of 1956, I did odd jobs working as bricklayer on several different job sites, including bricklaying on a bowling alley renovation. At the end of the three-week project, my father got me a job on a brick loadbearing apartment house on upper Broadway near Van Cortland Park. It was with Isadore Rosen and Sons, who were lumpers. A lumper is a contractor who does everything on the lump sum price. It was tough, it was brutal, and it was summer. I noticed every morning and every afternoon that the foreman gave people not keeping up, an envelope and they were fired. So the practice was to have company men raising the corner so that they could raise the line faster and force everyone on the line to keep up with progress. At 3:30 p.m. the first day, they announced that the patriarch, Isadore Rosen, died and we all got Tuesday off without pay.

Very benevolent! We worked so fast with wet brick that I went through three pairs of work gloves. At the end of the first day, all 10 of my fingers were bleeding. I walked across the street, got in my car, and drove home. When I got home, I had excruciating pain when I washed my hands because there was no skin left on any of my fingers. The good news was I had a day off to recuperate. Wednesday morning, at around 11:00 a.m., I stopped and stretched because I was backing up and getting ahead of the old-timer and as I stood up and stretched they handed me an envelope. I asked if it was payday and they said no, I was fired. My first reaction was tears came down my cheek – it's not nice to be fired. I then walked across the street, looked back at the jobsite, and jumped with joy that I was free.

The next summer in 1957, I worked on the Chase Manhattan headquarters 60-story building. This was really a seminal year for me. It was during this job that I met Eugene Fulham, my first real serious engineer role model. He was a Manhattan graduate, who ran the day shift. In 1945, at age 17, he was a Marine on a troopship on his way to invade Japan. He came back home and on the G.I. Bill he attended Manhattan College and graduated from there in 1951. At age 29, he was in charge of the entire foundation construction for one of the tallest buildings in New York City. They would work around the clock. Another Manhattan graduate, Bill Solomon, who later became a lawyer, ran the night shift. I remember working six days a week. But it was George Sengelub, who was designing the timber trestle to bring trucks down into the hole, which was 80 feet deep, that I studied with great interest. I realized that structural engineering was what I wanted to do. George was using a slide rule calculating timber members that would support the trucks going down the ramp. I started asking questions about how he was doing it and had the know-how to do it. I was thoroughly impressed. One of my jobs was to measure the perimeter cofferdams, including the length, width, and height. This was to determine the number of cubic yards of concrete that would go into each cofferdam. I remember climbing 80 feet up and down on rebar dowels sticking out of the side of the prior pour. The rebar dowels were offset. I would just swing on the bar above and chin up to continue my ascendancy to the top, rather than waste my time taking the stairs. During the pour, I would check in the concrete trucks right next to the Federal Reserve building and check that they were not mixing for too long in order to be sure that the quality of the concrete would remain valid. This was the summer between Iona Prep and Manhattan College.

In 1958, I worked for Thomas Crimmins contractors on 9th Avenue and 30th Street, building an open-cut tunnel. My job was to survey an underground tunnel on Ninth Avenue, between the General Post Office and the Morgan Annex. In 1959, I worked as a surveyor on a very tall steel framed building – the First National City Bank Building on Park Avenue. Ken Borst, who married my cousin Jane Thornton, both Cornell civil engineering graduates, was with George A. Fuller, the largest contractor of the day, who built the Flat Iron Building. Ken moved to New Haven to be in charge of a Yale University project and ultimately ended up being director of facilities for Yale University. He later hired my firm to solve a bunch of problems over the years. My immediate boss on this 40-story steel frame building was Tom Durant, project manager and a Cooper Union graduate, and Larry Maletta, head of survey teams. Ironically, Larry later turned up at One Tampa City Center as the superintendent with Pavarini Construction.

My first day on the job they showed me that if I was walking high steel, I should always carry the surveying instrument toward the outside in case I fell, so I would only be down two stories instead of 40 stories. This was very reassuring. I spent the whole summer walking high steel surveying four-foot marks and measuring column splices. I learned that the east side of the building warms up first when the sun comes up, which makes the column splices higher on the east and south sides. As the day goes on, the column splices get higher on west and the east splices get shorter. I had to take this into account as I surveyed the elevations of the column splices – as it was a moving target all day.

I learned to duck-walk the beams, putting my feet on the bottom flange and sliding my bottom across the top of the beam. I could walk high steel, two stories up, but anything more than that, I froze – acrophobia. Although the ironworkers laughed at me, at least I lived. The seat of my khakis would constantly be black at the end of each day. This was not my favorite summer job. On my way to work, I would fear that I might not make it through the day. This is probably why I became an engineer who could spend more time in the office. Later in life, my family shared with me the famous photos, *Lunch atop a Skyscraper,* of several men eating lunch on a beam of the RCA Building overlooking the New York City skyline. None of the men had safety apparatus. The photo was taken in 1932 during the Great Depression. My family shares the belief that five of the men in this photo were my Irish forebearers, Eddie, Les, Hugh (Babe), Paul and Bill Thornton. This makes sense when thinking about the lineage of my family in masonry, construction, and engineering careers.

In 1960, I got a job with the New York District Corps of Engineers geotechnical division. I worked as a drill rig inspector on projects at West Point and ran calculations for levees and dikes along the Mohawk River in upstate New York. This was fun!

In the spring of 1961, Vincent DeSimone, from Manhattan College's Class of 1959, called my structures professor, Dr. Vincent Vitagliano, and asked if anyone was interested in a summer job doing structural engineering on flat plate concrete apartment houses. I raised my hand and the rest is history. I spent the whole summer doing structural calculations on concrete flat plate apartment houses and swore that this was something I would never do for the rest of my life. The name of the company was Lev Zetlin Associates. In 1977, the firm became Thornton Tomasetti.

Manhattan College & NYU

My older brother Bill is 18 months older than me and a really great guy, but a little bit of a pessimist. To this day, he still fears that on his way to the airport on a business trip, it's absolutely a given that something in the car will bust or break, that the car will get a flat, or that it will run out of gas. He admits to me that he just can't control his anxieties and cautious nature. So whenever we went on a family vacation, I would get in the car happy-go-lucky, smiling, and joking – with the Nike "Just do it" attitude. Bill would worry if we had air in the tires, oil in the crankcase, and gasoline. I guess based upon the cars my father had, he was probably right to worry.

Based on statistics, firstborns dominate as CEOs of large companies – the reason is they are risk-averse. Bill was not a risk-taker and I was. He graduated from Iona Prep in the top 10 percent of the class, got admitted to the engineering school at Manhattan College, and started college in September of 1957. During this time, I was a senior at Iona Prep goofing off and hanging out in the basement playing pool with all my friends. Bill, who would rather be chasing girls and fixing cars, was in engineering school because that's what my father wanted him to do. Looking back, it's so obvious to me that 17-year-old boys are very immature. Bill reached the end of his freshman year and he flunked out. He ended up going to summer school to repeat physics and calculus and got a D and an F respectively. They then kicked him out of college. This did not bode well for me, watching my smart brother being bounced from the school that I am about to enter for the same engineering program. I felt guilty that I was probably one of the reasons he failed, because I was always fooling around as a senior and he was distracted.

My father pulled some strings with the Archdiocese of New York and got Bill back into Manhattan College even though his grade point average was slightly under 2.0. When Bill went for an interview with the Dean of Engineering, the Dean said to my brother, "Thornton, you will never be an engineer." At that point, my brother Bill said under his breath, "You just watch." After his hardship freshman year, Bill ended up with straight As in junior and senior year. He went on to Case Institute of Technology in Cleveland for graduate and doctorate studies in applied mechanics. His doctoral dissertation for NASA was entitled, "Non-linear analysis of an Ablating Nosecone Reentering the Atmosphere" and involved almost all calculus. He then taught at Clarkson University in Potsdam, NY, for 12 years before entering the steel industry and becoming chief engineer of Cives Steel Company.

To this day, whenever I get an opportunity to speak to engineering deans and faculty, I try to make the point that bad calculus teachers can eliminate many great future engineers in their freshman year of college. There is finally a trend today toward getting rid of the "look left or look right" lecture, which tells students to look left and look right as half of them will not be there in the sophomore year. Schools make a big investment in enrollment today and try to keep the students whom they put the effort into recruiting.

Bill's college experience was a real lesson for me as I entered Manhattan College in September 1957. Freshman year in engineering school is boot camp. It is very difficult to take 20 to 21 credits a semester in order to complete a total of 160 credits for a bachelor's degree. Manhattan College accepted 200 freshmen knowing they only had room for 100 sophomores. My freshman year, I followed the pattern of my brother and got five Cs and a B, which got me a 2.2 index. Unlike Bill, however, I got Mr. Marano for calculus – a very good teacher. The next three semesters, I continued to improve and by first semester my junior year, I had a 3.6 index, five As and one B.

Dr. Donald O'Connor, Professor of Civil Engineering at Manhattan College, was the best teacher I ever had. He was the first teacher in junior year to truly explain why engineers need to study calculus. He gave applied, concrete examples in civil engineering and environmental engineering which made it obvious why we study calculus. Most other college calculus professors don't have a clue as to how to answer the question, "Why do we study calculus?" They make it very difficult. Dr. O'Connor's courses were a combination of technical and spiritual, morality and ethics lessons. He taught ethics and honesty and how to succeed in life. Every time I left Dr. O'Connor's class, I was inspired. His entrepreneurial spirit and his ability to attain large contracts within water treatment or river pollution, as well as a whole range of sanitary and environmental engineering, were inspirational.

Dr. O'Connor was the first person to tell me that I should aspire to become a member of the National Academy of Engineering (NAE). I worked for Dr. O'Connor and his associates during my senior year. After graduating, I became an honorary member of the Plumbers, which was kind of a social drinking group of environmental engineers. I followed Dr. O'Connor's track. He graduated from Manhattan in about 1950 and received his master's and doctorate degrees from NYU in environmental engineering. I maintained a 3.6 grade point average for both my junior and senior years and got accepted to Columbia University and NYU for my graduate work. After he started teaching at Manhattan College, Dr. O'Connor began several companies, which became very successful and profitable companies, with Wesley Eckenfelder, another Manhattan professor.

In high school I was always shy. I was tall, really skinny, and had acne. I thought girls would never like me. One of the most important things that happened in my life was being invited to pledge for Phi Kappa Theta, a national Catholic social fraternity. My Aunt Eleanor was Catholic and had influenced my mother, who was a Lutheran, to have us attend Holy Cross Catholic Church when we were young. Bill and I would walk alone to attend church every Sunday. The nuns insisted at our CCD meetings on Tuesdays after school that we attend church on Sundays. So, we went.

I pledged Phi Kappa Theta in the second half of my sophomore year, going through a six-week pledge. Pledging was a very interesting psychological experience for me – I was convinced I was going to get blackballed. The pledge process taught me a lot and probably contributed significantly to my maturation process. During the six-week pledge period, I was forced to become a leader. I was forced to come out of my shell as I chased fraternity brothers to interview them to get to know them. It changed my life and I learned a lot. The experience probably put me on the track to be successful in business and life.

Immediately upon joining the fraternity, our senior and junior fraternity brothers sat down with the new brothers and told us we had to run for elected offices in the American Society of Civil Engineers Student Chapter (ASCE), the Society of American Military Engineers (SAME), and Chi Epsilon, the civil engineering fraternity. We were told we needed to impress the faculty and the administration with our extracurricular activities to regain the fraternity's reputation since the college administration didn't like any fraternity. This launched me into a leadership role within civil engineering. I ultimately became an officer in ASCE, SAME, and Chi Epsilon. I had to run meetings, invite speakers and speak in front of my classmates, faculty, and guests. As an officer of ASCE, I met one of the most important people in my life—Arthur J. Fox, the editor of "Engineering News Record" (ENR), the Bible of the construction industry. Art was a Manhattan College graduate in civil engineering and the associate contact member for the student chapter of the American Society of Civil Engineers (ASCE) at Manhattan College. He mentored us on how to run the chapter and many other important leadership traits. As a result of getting into the fraternity, I ran for office for the student chapter. Approximately four times a year, we would meet with Art Fox in his office at the McGraw-Hill building on West 42nd Street.

Art is a great man and a great mentor. He taught us how to run a student chapter, how to do fundraising, and how to invite the best and brightest people in the New York area and construction industry to come to the campus to do presentations to the Manhattan College engineering students. I have maintained my relationship with Art Fox since 1959.

Art was also the brains behind the "Engineering News Record" Man of the Year awards dinner. This event later became The Award of Excellence Dinner, which is now held at the Marriott Marquis and attended by 1300 people. It is the event of the year to network and meet the movers and shakers of the engineering industry. "Engineering News Record" names

25 newsmakers every year. I received the newsmaker award four times – first, in 1978 for the Hartford Coliseum roof investigation; second, for the collaboration with Helmuth Jahn for the United Airlines terminal; third, for the Petronas Towers in Malaysia; and fourth, for being the founder and leader of the ACE Mentor Program. I don't believe anyone else has ever been a recipient of this "Engineering News Record" award four times. The last time led to me becoming the "Engineering News Record" Award of Excellence Winner in 2001. Art and a colleague started the Construction Industry President's Forum, which in 1998 became the Construction Industry Round Table (CIRT). Over 50 percent of CIRT members are national supporters of the ACE Mentor Program. Many of them are on the ACE national board. The relationships started with Art and "Engineering News Record's" support of the ACE Mentor Program, which places Art Fox among my very best friends and supporters. As far as I am concerned, Art Fox is God.

Through my fraternity, I met men who are still my friends for life. Erwin Gus Fruh (deceased), Michael Bellanca, Jack McNamara, John McCabe, Tom McGoldrick, Sal Monte, Ronnie Morgan (deceased), Tom Mulvaney, Bill Mullen, and many others. One of my classmates, Angelo Tomasetti, was the older brother of Richard Tomasetti, who would later be my business partner. Since graduating from Manhattan College, the third weekend in May each year is reserved for a weekend gathering of socializing and golf in the Catskills for 50 fraternity brothers and their spouses. It's miraculous that of the 50 fraternity brothers, there have only been about three divorces and a couple of widow/widowers. This is contrary to the national norm, I imagine in part because Phi Kappa Theta was a national Catholic fraternity. The other reason for so few divorces is that the people who were selected to pledge were extremely adept at marrying really dynamic women. We attended each other's weddings, including Mike Bellanca and his wife Jan's wedding in Roanoke, VA, and Gus Fruh and his wife Gwen's wedding in Sheboygan, WI. We had little money in those days and had memorable bus, car, and train trips to see one another.

All through college, since we didn't have a fraternity house on campus, our group would have parties every Saturday night at our parents' homes. The theory was if we introduced the guys we hung out with to our parents, they wouldn't give us a hard time. Phi Kappa Theta was really a turning point in my social life. It was during this period that I blossomed into a leader and ran the Manhattan College Green and White Ball, a major dinner dance at the Essex House Hotel on Central Park South, and the April In Paris Ball, another event in Manhattan. I also started attending Manhattan College Holy Communion breakfasts at the Commodore Hotel on Lexington Avenue and 42nd Street, which is now Trump's Hyatt, right in Grand Central Station. I started seriously dating Patricia Podaski, who fit in perfectly with my fraternity brothers and their dates. During this time, Patricia's health seemed to be stable. We enjoyed being social and attending fraternity events.

3
Starting a Family

Married Life

I MARRIED PATRICIA PODASKI ON September 9, 1961 at the St. Simon's Stock Church in the West Bronx. Our reception was at Alex and Henry's, an Italian "marriage mill" restaurant near Yankee Stadium. On our wedding night, we stayed at my family's farm in Cairo, New York. We honeymooned in northern Maine, selecting Maine because it was all a starving graduate student could afford. We stayed at a nice resort in the Rangeley Lakes Region. I got up the next morning and dove into the lake and I couldn't find my you-know-what's because it was so cold. We stopped in Boston on the way back and saw the movie, "The Guns of Navarone."

When we got home from our honeymoon, Patricia went back to work as a bank teller at Chase Manhattan Bank and I started graduate school, which wasn't that challenging, so I went back to work part-time for Lev Zetlin. Our one-bedroom apartment was on the sixth floor of a six-story walk-up at 2108 Walton Avenue at 181st Street and Walton Avenue in the Bronx. Although it was a convenient location to NYU where I was attending graduate school, only four blocks up the hill, it had a roach-infested kitchen. I could actually walk into the kitchen at night and hear the roaches making love on top of the kitchen cabinets. If anyone exterminated their apartment, all the roaches went to the apartment next door. It was great fun.

Patricia and I wanted a baby right away and October seemed to be the most fertile month. Diana, our first child was conceived that year and I literally had to push my wife up the stairs during the last months of pregnancy. Diana Lynn was born on July 28, 1962. Katherine Ann, nicknamed Kathy, was born on July 25, 1964. Both daughters were born at Jewish Memorial Hospital on Sherman Avenue and Broadway in northern Manhattan. The obstetrician generally took the month of August off, so Patricia had labor induced the last Saturday of July for both Diana and Kathy. In those days because fathers were not invited into the delivery room, I usually took a couple of books and set up in Fort Tryon Park nearby. I just found a park bench near a pay phone, read a couple of books, and called the hospital every hour. I would just wait for the call that my child had been born.

Right after Diana was born in July of 1962, we moved across the street to a third-floor apartment at 2100 Walton Avenue, which made life a lot easier. My mother-in-law Anna Podaski and her husband Stanley were the superintendents and lived in the basement. We put a diaper bucket with diapers and laundry on the dumbwaiter and sent it down to the basement and it came back clean and ironed – even my underwear. Having a built-in-babysitter in the form of my mother-in-law was not a bad thing for a young couple that was looking to continue with a social life.

Soon after Diana was born, I finished my requirements for my master's degree. The chairman of the Engineering Department, Dr. James Michalos, encouraged me to continue and get my PhD, giving me a fellowship which would pay for 18 credits a semester. I finished my PhD coursework by June 1964, a month before my second daughter, Katherine Ann, named after my maternal grandmother, was born. I had bought a multiplex FM stereo for $200, which was a lot of money in 1964 when I was only making $100 a week. Kathy used to stand in front of the stereo and dance to "Somebody to Love?" by Grace Slick of Jefferson Airplane. I think she learned to walk to that song. I then decided to take, what was supposed to be a year off, to do my doctorate thesis. I had $5,000 in the bank, two children, and was paying $68 per month for a one-bedroom apartment. I was broke by October 1965 and returned to Lev Zetlin Associates (LZA).

A Growing Family

Although I was busy with LZA, I made it a point to always be home with the family on weekends. I had a car, which I tended to leave at my parents' house in the East Bronx. From 1964 to 1966, we generally headed up to the family farm near Cairo in the Catskills in the summers. We had 80 acres, with three quarters of a mile of frontage on Catskill Creek, with a deep swimming hole. The whole family was generally there including my brother Robert. Bill had already moved on to Cleveland to Case Institute of Technology to get a master's and doctorate degree. We grilled every weekend and spent time outside, which was a relief since our one bedroom apartment only had a window air conditioner, which made us feel claustrophobic. Other weekends, we went over to my parents' home in the East Bronx and freeloaded our lunch or dinner. They were thrilled to see the new grandchildren and we were thrilled to be fed.

Our third child, Charles Henry III, was born on June 3, 1967. From June to October 1967, we had three children and two adults living in a one-bedroom apartment. Although the diaper service from my mother-in-law continued, we started looking to buy a house. By paying only $68 a month rent, I was able to aggregate about a $6,000 down payment to buy a $27,000 house at 160 Lakeshore Drive on Bear Ridge Lake in Pleasantville, NY. It was a small three-bedroom house overlooking the lake that was quite clean – part of the New York City water supply system. It had a full basement, which I remodeled with the help of my friends Jack McKinney and John Kovac, and an open carport, which I enclosed to be a garage. Jack McKinney re-shingled the entire house with cedar shakes and added a formal dining room. The lake froze every winter and we skated almost every night. I had floodlights on the deck so we could see where we were going. In the summertime, we swam in the lake and enjoyed the beach.

After we bought the house at 160 Lakeshore Drive, our families started visiting us on weekends. My brother-in-law, Ray Podaski, was into fabulous rock music and introduced me to some really great bands and groups. From 1961 through 1967, I started mentoring him and got him into Bronx Community College for pre-engineering. He worked at the Joint Venture office for the American Airlines Hangar project. He was really respected by Joe Denny and Joe Thelen and all the team members on the American Airlines Hangar. He didn't like living at home in a one-bedroom apartment with his parents, so in the guise of attending a wedding in Denver, CO, he left New York and never came back. He ended up getting married and never went back to college.

We joined Welcome Wagon Newcomers and made a lot of new friends which led to a couples' gourmet group that met monthly. Some of our best friends came out of that association, including Richard and Jeannie Pactor, Les and Nancy Mayer, Steve and Iris Schwartz, Steve and Lissa Wesley, and Richard and Angela Doesher. The Pactors had two children, Philip and Susie, who became instant friends to Diana, Kathy, and Charlie. We took family vacations with the Pactors to Lake Winnipesaukee in New Hampshire and we all spent a lot of good times together.

In 1970, after my first trip to Europe with my wife, we brought back Austrian knapsacks for each of the children. We lived at the foot of Bear Ridge Hill, which remained undeveloped in 1970. On Saturdays and Sundays, I would load up the children's little packs with Sterno stoves, canteens, cans of Campbell's soup, and matches, and we would hike up the hill and find a comfortable spot and cook lunch. The kids loved it. On occasion, we would head up to the Hudson River Valley and climb Breakneck Ridge, which is a spectacular elevated ridge overlooking West Point.

When Charlie was eight in 1975, I volunteered to be the cub master for an existing cub pack at Holy Innocents Church in Pleasantville. During the Cub Scout days I had started to get involved in canoeing and I took the cub pack canoeing on the Mullica River in the New Jersey Pine Barrens. This was a great trip for the boys and the river was shallow, so safety was not an issue. Through my leadership over the next three years, I grew the pack to over 125 boys – it became one of the most successful packs in Westchester County. Between 1976 and 1977, when the Arab oil embargo was going on, we met in the community hall of the Holy Innocents Catholic Church. One day, the pastor of the church called me and told me that we were no longer welcome to use the hall because of the cost to heat it. He couldn't afford to heat the hall and I asked him what I should do with 125 kids. He basically said that it was my problem. This was the beginning of my deepening lack of respect for organized religion. This was a great way to drive away parishioners. I immediately drove over to Emmanuel Lutheran Church on Manville Road in Pleasantville and introduced myself to the pastor and described my cub pack. The pastor at the church said to bring all the kids there to meet and that it didn't make any difference what their religion was. We met at this church until I stepped down as cub master.

As Diana reached the age to join the Girl Scouts, my wife became the Girl Scout leader. She loved it and Diana stayed in the Girl Scout program, achieving the rank equivalent to Eagle in the Boy Scouts. Next, I tried to convince the Girl Scouts that my neighbor, George Cangelosi, whose daughter, Donna, was in the Girl Scout group with Diana, and I should take the girls on a camping trip. I was really taken aback when the mothers said no. I was persistent and the following year George and I took the girls canoeing in the Pine Barrens of New Jersey. It was a great trip. We got to the campsite a little late and all of the sites except one were occupied by Boy Scouts. Although the mothers may have questioned how George and I could keep track of adolescent girls on a campout, the real threat to their little sweetie pies was the Boy Scouts. Some of the Boy Scout troops were actually interested in girls and some of them were just interested in beating up on girls and tearing down their tents. George and I stayed up all night protecting our campsite from these villainous boys.

During this time, Patricia continued to get radiation treatments for her lymph nodes. She had been diagnosed with Hodgkin's lymphoma, but she wouldn't let me tell anybody about this, including the children. She wanted to be perceived as strong and healthy and certainly didn't want anyone feeling sorry for her. I suspect, however, that the children had an inkling that their mom was not in perfect health.

By the time my kids were between four and seven-years-old, we became more affluent. Patricia and I did not want to spoil them, so I gave them jobs. I made them sign contracts and I made them write out and send me invoices. When I paid them, I insisted that they send me a thank you note. This was my way of showing them how a business operated and how things were in life. Kathy and Diana used to shine my shoes and send me invoices and thank you notes. Kathy went on to become a CPA. I remember reading an article in the Allegheny Airlines in-flight magazine written by a father as to why you should not buy your daughter a horse. Diana rode horses at the Bear Ridge Stables and she begged me for a horse. I said no, but in order to make it fair, we had a family vote about whether to buy a horse or a swimming pool. The family voted for a swimming pool. Charlie's jobs were cleaning the pool and mowing the lawn. He still has the accounts receivable from his invoices and often tells me that the experience helped him learn the importance of responsibility and finishing tasks you are assigned. Today, Charlie has a successful career in sports construction facilities management.

Fun Family Times

By 1973, Gable Industries had acquired LZA and my income had increased significantly. We bought a 100-year-old converted barn at 324 Lake Street in Armonk, NY – the home of IBM. Armonk was in the township of

Mount Pleasant, the same as Pleasantville, so there wasn't much of a change except schools in Armonk were significantly better than in Pleasantville. It was a great house in which to raise three great kids, plus it had room for an additional three senior citizens when the time came. The house cost $73,000, which was a stretch, but it was worth it. It had about 5,000 square feet. Its features included a full basement with a huge fireplace and a nine-foot ceiling. There were two bedrooms, a kitchen, bath, living room, and dining room on the ground floor with two bedrooms on the upper floor, plus a huge playroom with a cathedral ceiling. On the upper level, there was a 750 square-foot sleeping loft, which is where Diana and Kathy hung out. At the time, it was thought that Diana was the slob. It was obvious when she left for Gettysburg College, however, that it was Katherine who was the messy one. The oldest always gets blamed for everything.

Between 1968 and 1972, I traveled extensively during the American Airlines Super Hangar project. It was a short ride to the railroad station and a 45-minute train ride to Grand Central Station. All of our offices in those days were within walking distance of Grand Central Station and less than a one-hour drive to JFK and the La Guardia Airports, so traveling wasn't that difficult. I remember one day when I returned from one of those trips to California, Patricia had lined up the three children against the front wall. When I walked in, the kids asked their mommy, "Who is that man?" I got the message. I resisted any travel that kept me away over weekends. The only time I stayed away over weekends was during my trips to Moscow in October 1976, when I went twice and stayed two weeks. The rest of the team actually stayed from Labor Day through Christmas. There was no way that I would do that.

Canoeing and Whitewater Rafting

By the early 70s, my brother Bill was on the faculty at Clarkson University in Potsdam, NY. He was on the faculty for about 12 years teaching structural engineering. Clarkson is a rather inexpensive place to live, so he bought a camp on Higley Flow. His summer place was 10 miles from his main domicile and Higley Flow was on the Racquette River. This was the beginning of our whitewater canoeing career in the Adirondacks.

Sometime around 1974 I started canoeing. I purchased a 17-foot Grumman aluminum sailing canoe and a 17-foot ABS Blue Hole white water canoe. I canoed the Croton River, between the dam and the Hudson in the Croton Gorge and canoe sailed the Hudson River near Croton Point. I also took the canoe up to Lake Winnipesaukee in New Hampshire and sailed all around the lake and on the Housatonic River in Kent, CT. But the best canoeing was in the northern Adirondacks.

Brother Bill had access to canoes from Clarkson University, so we started researching canoe trips. My family would drive up with our canoes on the roof, meet Bill and his family at Higley Flow, and then the spouses would drive us to our put-in point. We would canoe with the older kids. On the early trips, we left the younger ones behind, and the spouses would pick us up at a pre-determined location. In 1975, we started with a flat water trip from Axton Landing to the village of Tupper Lake. This was the trip across Tupper Lake with a stiff head wind and water about three feet deep with lots of stumps below us. The natural level of Tupper Lake was raised for hydroelectric generation, and to assist the transport of logs down the Racquette River to the St. Lawrence.

This was not much fun and a lot of work. So we advanced to Class 2, 3, and 4 rapids quickly. We would do two-night and three-day trips. We canoed all across Tupper Lake, into the Racquette down to Moosehead Rapids. The first trip down Moosehead Rapids, which is several miles long, is actually rated a Class 4 due to its lack of accessibility, standing waves and souse holes. We ran Moosehead Rapids twice. The first time we did it, we backpacked all our stuff up along the east shore and tied Bill's dog, Blackmale, to a tree, hiked back to the canoes, and ran the rapids. In retrospect, we were in over our heads. My younger brother Robert and daughter Kathy rolled the canoe over in the middle of the rapid and we couldn't see them. Charlie was crying and we were all very concerned. Blackmale broke loose and swam out into the river and we caught him on a flat rock. Finally, the canoe came down to us and Robert and Kathy were able to climb out on the west shore and come down and join us. We continued down the river to Carry Falls Reservoir, canoed the entire length of Carry Falls and met the spouses at the dam.

In 1981, we did Moosehead Rapids again. We did not backpack our gear around the rapid. We blew right through the rapids in three canoes as if we knew what we were doing. We always canoed in late August, because by then it was cool and there were no flies or bugs. We would light a campfire every night. I really enjoy the beauty of canoe camping. You can bring coolers, stoves and many more things without straining your back. We also canoed the Oswegatchie River from Wanakena to High Falls and return in the Adirondacks. This was all flat water paddling.

I still love whitewater canoeing, but whitewater rafting is even better. It is one of the most exhilarating trips that I have done. I learned to love this on the Outward Bound trip and through Cataract Canyon on the Colorado River. Bill and his family and our family also whitewater rafted the Ocowee River in North Georgia. When my son Charlie got married in Colorado, anyone who wanted to went up to the Arkansas River with guides and we whitewater rafted the Arkansas River. It was unbelievably beautiful and exciting.

Life is a Bell-Shaped Curve

One of the more interesting people that I have run into in my life is Dr. Robert "Bob" Tener. Bob was the first Executive Director of the Charles Pankow Foundation. He held that position for approximately the first five years of the foundation's existence. I was pleased to be on the board of advisors for two terms at its startup. It was from Bob that I learned a lot about Lean Six Sigma, a methodology used by companies to improve business processes by utilizing statistical analysis rather than guesswork. Learning these management improvement strategies helped me later in forming new companies. One of Bob's favorite quotes is, "All of life is a bell-shaped curve." If one ponders this statement—it is so true. Life is a roller coaster with its ups and downs or steep inclines and steep declines. I was entering into a steep decline before a ride to the top.

About every five years, my wife Patricia would have a recurrence of the Hodgkin's lymphoma. From the time Diana was about ten-years-old, Patricia seemed to be sick more often, but she kept her illness fairly private. She would go in for some quick treatments and it seemed to go away. But in 1978, things worsened and by October 1978, she was admitted to Sloan-Kettering Medical Center. The doctor who cared for her was actually the same doctor treating the Shah of Iran's cancer. It came close to mid-November and I was anxious to bring Patricia home for Thanksgiving. I went up to Sloan-Kettering and met with the doctor. He said, "Sure you can take her home, but don't bring her back as there is nothing more we can do for her." I said, "What? Do you mean she's going to die?" He said she was going to die. I was shattered. I didn't realize how far her illness had progressed. I sat in shock and then walked back to Grand Central Station. On the way, I called Richard Tomasetti, my closest friend, and told him what was going on.

I brought Patricia home on the Wednesday before Thanksgiving and we were able to have Thanksgiving dinner, even though she was in a wheelchair. Because my father-in-law had died two years before, my mother-in-law, Anna Podaski, had moved in and was caring for Patricia while I was working. The Monday after Thanksgiving, while in Connecticut at meetings, I called home to see how Patricia was. My mother-in-law said things were not going well. I then called Steve Wesley, one of my medical doctor friends, and he called the ambulance.

The kids were asked to go upstairs while they took Patricia out of the house in the ambulance. We had protected them from knowing about their mother's illness. My mother-in-law rode in the ambulance as they took

Patricia to the hospital. I followed in a car and by the time we arrived at Westchester Medical Center, Steve Wesley came to me and said, "I'm sorry, but she's gone."

Patricia and I had been married 17 years and they were good ones. You never fully appreciate what you have until you lose it. She was my partner, my friend, and a fabulous mom.

My mother-in-law and I got home that night and the children were actually sleeping. The next morning, I went in and told the children that their mother was no longer with us. The children seemed to cope with their mother's death fairly well. They said later in life that the extended family's care, especially the care provided by their grandparents, comforted them. I remember my doctor friends encouraging me to get counseling after Patricia's death, but I said I didn't need it because I had my family to help me work it all out.

Years later, Diana shared with me the important role a guidance counselor played in her grieving after her mother died. Since Patricia was sick for a number of years before her death, Diana had assumed the role of caregiver for her brother and sister. Even though the grandparents were playing a more prominent role in the care of their grandchildren, Diana still felt responsible for her siblings. During Diana's senior year of high school, soon after her mother had died, a high school guidance counselor sensed that Diana may not want to attend college because of her sense of responsibility to her siblings and asked her to come into his office to discuss this. At the time, Diana was very upset with the counselor for doing this. The next fall, however, Diana enrolled at Gettysburg College. Recently, she contacted her guidance counselor from high school and thanked him for being there when her mother died and encouraging her to attend college.

Diana and Kathy later shared with me that neither of them really dealt with their mother's death until they were adults with their own children. For Kathy, the healing occurred years later when she experienced the death of Patricia's best friend, Jeannie Pactor.

Holding it Together

There were several things I needed to do immediately upon the death of my wife in November 1978. First, I had to find a way to keep my family intact. Diana was 16, Kathy was 14 and Charlie was 11. My mother-in-law, Anna Podaski, and my brother-in-law, Ray Podaski, would generally be at my home on 324 Lake Street on most weekends. The children loved their Uncle Ray, who they called Ray Ray. I eventually asked my mother-in-law to retire from the New York City Welfare Department and move in with me. I needed help. With her usual astute manner, she said, "Charlie, you're a young man. At 38, you will meet somebody and you will get remarried." I told her not to worry about that at the time, but instead to cross the bridge that was right in front of us and get on with a continued family unit.

In late 1978, Anna moved in. She was a nurturing and loving grandmother. This enabled me to continue working full-time. Richard and I had just bought the company and I had no choice.

At this point, my parents were nomadic and pretty much spent winters in California with Aunt Eleanor and summers in the Catskills. They would visit my home when coming and going. Right after Patricia died, they moved in for a short time to help the transition. Sadly, shortly thereafter, probably in the summer of 1979 my father had a debilitating stroke. He was building an extension to the country home in the Catskills by himself when it happened. Apparently, he had unilaterally decided to stop taking his high blood pressure medicine. My mother called and told me that my father had had a stroke and was in Albany Hospital. She said he was kind of out of it, so we drove up. It was really a tough situation to see a very strong father in that condition. My mother said she was going to put him in some kind of a rehabilitation clinic, but I refused to let this happen and they both moved into 324 Lake Street. There was plenty of room.

This was the beginning of the extended family living together, which actually lasted all the way through the 80s and most of the 90s. When I tell people that for approximately 15 years, my mother and father and first mother-in-law lived together happily ever after – most people can't believe it. Each grandmother took a different role. When my mother moved in, she took the role of laborer – cooking meals, taking care of the house, and transporting the children. This was a very natural progression for me as I was raised with the extended family concept. While it may have been a risk to have the grandparents take such a key role in raising my children, I think it helped the children overcome this hardship in their lives. I was proud of their resilience and ability to adapt. I believe it made them stronger people.

After Patricia's death, there were two more things that I had to do – to start skiing and start sailing. These are two sports that I knew my children would continue to do with me, as they got older. I could invite them to go skiing in Steamboat Springs, Vail, and Breckenridge, or to go sailing on a chartered 50-foot boat in the British Virgin Islands. They were always delighted and always asked if they could bring a friend.

During the Christmas holiday in 1978, I took my kids to my brother Bill and his wife Joan's place in Potsdam, NY. We skied at a Clarkson ski area, south of Potsdam. This was an escape from the surroundings that reminded all of us of Patricia.

Philip and Suzy Pactor, Richard and Jeannie Pactor's children, went to school with Diana, Kathy, and Charlie. The Thorntons and Pactors generally took an annual vacation every year to places like Québec and New Hampshire. Richard Pactor and I were pretty good cross-country skiers and we used to go skiing on the golf courses in Westchester County. Our family cross-country skiing trips generally entailed driving to southern Vermont and staying in cross-country ski resorts. Although it was not fancy, we got fantastic exercise. Unfortunately, Diana went on a high school bus trip to Catamount in the Berkshires to go downhill skiing. Upon her return, she announced that cross-country skiing sucked, it was too much work and she didn't want to do it anymore.

So we started skiing in the Berkshires in western Massachusetts and met Brian Fairbanks, the General Manager of Jiminy Peak. Brian was an affable young man with a young family and he told us that he intended to build condos at the base of the mountain. He needed buyers to self-finance. Richard Pactor and I met with Peter De Gaetano, a lawyer friend, and he told us we were out of our mind if we put money down and the project stalled. He said we would never get our money back. The problem with lawyers is they always try to save you from yourself. They give you the right advice, but you end up doing what you feel is right. We purchased a three-bedroom condominium at Jiminy Peak with Richard and Jeannie Pactor and we started downhill skiing. Then when Richard Tomasetti and Bob Liss came up to go skiing, they decided that we would buy another three-bedroom condominium, with a three way split between Liss, Tomasetti, and Thornton.

Later, Richard Tomasetti and I bought a studio at the Lodge at the foot of the mountain. Most weekends, Bob and his wife, Rochelle, and their son, Michael, and Richard and his two daughters, Yvette and Denise, and friends would go skiing. At the same time Jay Prasad and Dan Cuoco also bought a condominium. These condominiums later became the base for the company hikes. We had just the right group for both skiing and socializing. This is really why sailing and skiing have continued to play such a large role in my life. Not only were the condos in the Berkshires great for family togetherness, we had many clients come up and stay.

Nearby was Brodie Mountain, which was for the singles and drinking crowd. After skiing, I would drive to Williamstown to visit the Clark Art Museum and Williams College Museum of Art where I developed an appreciation for French Impressionist painting. The Berkshires are a place filled with many cultural events. We generally rented the condos out for the summer when Tanglewood and Shakespeare on the Mount and other cultural attractions were in play. The summer was a time for sailing.

Going back in time, my wife Patricia and I were invited to spend a week in Antigua at the Halcyon Cove Hotel complex in Dickinson's Bay on the north side of the island. We were with neighbors, John and Sarah Petriello, from Lakeshore Drive. One day we went down to Nelson's Dockyard and

Shirley Heights. We were standing on the top of Shirley Heights, which is probably about 800 feet high with spectacular views of the Caribbean, and I watched a boat on a beam reach head out to the south in 30 to 35 knot winds and huge waves and whitecaps. At that point, I said to myself that I was going to do that someday.

In 1979, sailing became a terrific pastime for family togetherness, as well as client sailing. In January of that year, I called Bill Farr of Fairwind Yachts, a yacht charter company in Greenport, Long Island, and I chartered a Pearson 35-foot sloop for five days in June. This was right about the time the gas lines were all forming.

Diana, Brooks Hayes, and I took the Coast Guard Auxiliary sailing course for two nights a week for 10 weeks. Brooks Hayes was the son of a client of mine, Tom Hayes at Cushman and Wakefield. There was no way that I was going to allow either Diana or Brooks to get a higher grade than me and they didn't. I guess I am competitive. In June, I chartered the 35-foot sloop and took Diana, Kathy, Charlie, Brooks Hayes, and Susie Pactor and left for Greenport. When I chartered the boat, Bill asked me if I knew how to sail. Of course I answered "yes!" He then asked me what I sailed, and I said a 17-foot Grumman sailing canoe. He told me if I could sail that, then I could sail anything.

Although I had been on boats since I was eight-years-old, I did ask to have a professional captain join us for the first two days. Bill agreed and the first two days we sailed all around Shelter Island, one of the greatest sailing places on the East Coast. Here you can experience every single condition you would ever want to encounter. At the end of the second day of sailing, Bill said that we were ready and to go for it. We sailed to Three Mile Harbor, Montauk and back to Greenport, CT. It was great and it really whetted my appetite for sailing.

4
Founding Thornton Tomasetti

Lev Zetlin Associates and PhD

LEV ZETLIN WAS BORN IN Baku, Russia. His father was a white Russian Jew who had moved the family to Tehran when the Russian Revolution was won by the Reds. He went to high school in Tehran, engineering school at the Technion in Tel Aviv, Israel, and was involved with the British Palestinian Army during World War II. He spent most of the war interrogating German officers in prisoner of war camps in Italy. This produced a man who had unbelievable psychological abilities, specifically the skill to convince people and motivate them. After fighting in Israel's war of independence and serving as chief engineer for civil works for the Israeli Air Force, Lev came to the United States to attend Cornell University, where he received his master's degree and PhD in structural engineering in the early 50s. He came to work in New York for the firm Ammann and Whitney, where he designed some cable-stayed folded plate hangars at John F. Kennedy International Airport. He didn't like working for anyone else, so he started his own firm sometime in the mid-50s and taught at Manhattan College to make ends meet. Most of the key staff in the office when I joined the firm were Manhattan College graduates – James

Chaplin, Vincent DeSimone, George Feddish, Frank Marino, and Mike DeLouise.

Dr. Lev Zetlin was a charismatic, creative structural engineer with strong abilities in innovating and in convincing clients to try new things. He surrounded himself with very bright, young, well-educated engineers. Once he liked you, the first thing he would do was give you a copy of Machiavelli's *The Prince*. He would suggest that you read it and then have a discussion with him. It is a great book and actually still applies to life and business. Things haven't changed since Machiavelli wrote it in 1532. Lev Zetlin's first invention was the bicycle wheel roof, which was used on the Utica Municipal Auditorium in Utica, New York in the late 50s. In 1962, his startup company landed 14 contracts for the New York World's Fair, which would occur in 1964 and 1965. Considering that the firm had a staff of about 10 people, this was an unbelievable accomplishment.

Based upon my W-2 form in 1964, I averaged about 35 hours a week with LZA and was a full-time graduate student. Vincent DeSimone was essentially running the office and James Chaplin was the best structural engineer in the office. Early in my career, Jim Chaplin was the most positive influence on my ability to become a really good structural engineer. Jim had the uncanny ability to cut back a structure in the pre-computer days and solve highly redundant statically indeterminate structures with great finesse, talent, and creativity. These structures had a greater ability to withstand collapse resulting from the loss of a single member.

During the first couple of weeks at NYU, I met Harry Armen, who was also beginning his master's program. My wife and I became very good friends with Harry and his wife Mary, which led to a lifelong friendship which continues to this day. The first three years of my graduate program at NYU, while working full-time for LZA, I was supposed to work 20 hours a week as a graduate assistant. I interacted with two great administrative staff members of NYU's Civil Engineering Department – Mrs. Spanier and Mrs. Altman. As it turned out, the department didn't have much for me to do, which enabled me to ramp up my working hours at LZA. One day in the middle of my second year, these two women came to me and expressed concern that Tony, a graduate assistant who was one year behind me, was moonlighting when he was supposed to be working at NYU. Because I was married and raising two children, these two women thought the world of me and had no idea that I was moonlighting too. This was hilarious to me.

During this time at LZA, while I was working on the Kline Biology Tower at Yale University with architect Philip Johnson, I was also working on a number of the World's Fair pavilions. These included the Eastman Kodak Pavilion, the New York State Pavilion, the Hawaiian Pavilion, the Mexican Pavilion, the Minnesota Pavilion, and the Spanish Pavilion. Although only out of college one year, I got the lead position on the Travelers Pavilion and had the pleasure of working with Jim Chaplin. While I was working on these projects, I was taking the core graduate courses in engineering mechanics, elastic stability, plates and shells, structural analysis, variational principles, elasticity, plasticity, nonlinear methods, and continuum mechanics. Having the opportunity to work on these unique projects afforded me a special opportunity to immediately apply what I was learning at NYU from one of the greatest teachers I ever had, Dr. Arnold Kerr. Most of these projects utilized structural solutions that had never been done before. Considering the firm did not have a computer, we used very interesting alternative approaches to perform the structural analysis of these structures. While many of the World's Fair projects were relatively straightforward, a number proposed interesting challenges.

I assisted Lev Zetlin and David Hoffman on the Eastman Kodak Pavilion, a free-form concrete shell. Since there was no mathematical equation to define the shell surface, we took the designer's model and had a Chicago-based firm build a micro-concrete model of the shell, install strain rosettes and linear strain gauges at the critical locations, vacuum load the shell and measure strains. We then converted by hand the strains into bending moments and in-plane forces and actually designed the reinforced concrete shell.

It was on the Eastman Kodak Pavilion that I met George Pavarini, one of the most fascinating men I have met in my career. He was a successful, self-made man, who grew up in New York City and attended Xavier High School, a Jesuit military school. In the middle of his high school program, the Japanese bombed Pearl Harbor and his high school years were accelerated. At about 17 years of age, he found himself in the Navy Air Force flying Chance Vought Corsair fighter planes in the Panhandle of Florida. Shortly thereafter, George found himself on a US carrier in the South Pacific providing air support for the United States Marines on the beach of Okinawa.

George returned to the United States and received an architectural degree from Rensselaer Polytechnic Institute in Troy, NY. He started his architectural firm in New York City called Belfatto Pavarini and then formed Pavarini Construction. George was an extraordinary construction company executive. As Lev Zetlin stood under the concrete shell with George Pavarini and his partner John Contegni and learned that the long span portion of the shell was continuing to deflect a little more than predicted by the calculations, George and John turned to Lev and asked him what he was going to do if it didn't stop deflecting. Lev reached into his pocket and pulled out a paper and said he had his tickets to Argentina. I thought that was hilarious.

Later, George Pavarini was also the constructor on the tallest building in Florida in 1981 – One Tampa City Center – the GT&E Corporate headquarters. I got to know George quite well at that time as he joined a few of my Male Bonding sailing trips and had a thoroughly good time. He is one of the best joke tellers I have ever known. He and Frank Marino are rated right at the top for joke telling.

I worked with Frank Marino and George Feddish on the New York State Pavilion. We retained a professor from Pratt to build a model of the elliptical bicycle wheel roof. In addition, Frank Marino developed a very clever influence line analysis to solve the structural design of the elliptical ring. This was one of the cleverest pieces of engineering that I had seen to date. For the red umbrella Travelers Insurance Pavilion, which was circular, we used a small bicycle wheel roof. A circular roof is much easier to analyze because all cable pairs are identical. To analyze and design the boomerang frames of a bicycle wheel roof, I used virtual work under the direction of Jim Chaplin. We were able to put in an equatorial cable around the widest portion of the Travelers red umbrella and post-tensioned the entire boomerang frame to reduce bending moments.

Lev Zetlin Associates incorporated several creative approaches to management. The "deep end of the pool" approach involved figuratively pushing young employees into the deep end of the pool – if they didn't come up, so what! This approach was not for the meek as it encouraged employees to take initiative, take chances, take controlled risks, and push the envelope. Today, when I talk with George Feddish, Frank Marino and other early LZA key staff, they can't believe that at the ages of 22 to 26 years of age, we all designed 14 World's Fair pavilions, at least five of which had structural systems which had never been done before.

The other approach to business while building the World's Fair pavilions was the "go down in victory approach." Our team would always propose three schemes for every project: the wild crazy scheme, the small step forward for mankind scheme, and the plain-vanilla scheme. Rarely did the wild and crazy scheme get selected, but that was the one the architects always loved and the one that left a lasting impression that we were very creative. For these reasons, we would propose the wild and crazy scheme and go down in victory, even though in reality, we usually ended up doing something between the wild and crazy scheme and the small step forward scheme. Most engineers spend their careers doing plain vanilla schemes.

While working on my master's thesis abstract, I was able to work in NYU's computer complex, which was the largest computer complex in the eastern part of the United States. This gave me a distinct advantage over everybody else at Lev Zetlin, as the company knew nothing about computers. Although Lev Zetlin got his PhD from Cornell in the early 50s, I knew about finite element modeling, so I was given all of the special assignments that involved theoretical and analytical skills. This enabled me to rise quickly through the ranks at the company. During this time, I felt like a nomadic tribesmen. Every week when I came back to work, the company was hiring so many people that my desk would be stripped of all of my work and I would find my papers stacked somewhere in the hallway, It would take me half an hour every Monday morning to regroup.

In 1963, when it became apparent that I was on the fast track at NYU to get not only my master's degree, but also a PhD in structural engineering, I approached Lev to tell him that I planned to resign and take a year and a half to two years off to do my PhD thesis. To my surprise and dismay, he tried to talk me out of it this. He was in the middle of the design of all of these pavilions at the World's Fair and he needed me. I was very disappointed with his self-serving attitude. Many of my colleagues at Thornton Tomasetti, including Richard Tomasetti and Tom Scarengello, all took coursework at Columbia and Polytechnic Institute of Brooklyn for their PhD's, but did not take a year off to write their dissertations. Despite his advice, I made my decision and I resigned from the company and took off two years to do my PhD thesis.

In June of 1964 when I left, I still had $5000 in the bank, but by October of 1965, I was broke with about $100 to my name and a wife and two kids to feed. I needed to go back to work. At that time, John F. Kennedy had said we would be on the moon by the end of the decade. In addition, National Defense Education Act Fellowships were available and the Vietnam War was starting to heat up. I received 14 job offers from aerospace firms, including Boeing, General Dynamics, Grumman Corporation, Martin Marietta, as well as Eastman Kodak for their lunar excursion module cameras. I decided, however, to go back to Lev Zetlin Associates because I thought I could do much better in private sector consulting engineering where one day I would own my own company. In late 1965, I went back to work at LZA, getting to work about 6 a.m. and leaving around 3:30 p.m. to go home and have dinner with the family before going up to NYU every night until midnight. I spent every Saturday and

Sunday all day with my thesis advisor. Finally, I had to approach my thesis advisor and tell him that if he didn't stop rewriting my thesis, I would die. He listened and I received my PhD in February of 1966.

As I started my career, I taught one night per week at Cooper Union, in the East Village neighborhood of Manhattan, to overcome stage fright of speaking to large groups. As time went by, it became apparent to me that the way to recruit the best and brightest out of the best engineering schools was to teach a senior class usually called the Capstone Course. Capstone indicates that it is a course taught by an outsider from the professional world and not a tenured professor. As a transitional course it introduces soon-to-be graduates to what the real world is all about. Because, over the years, I taught Capstone Courses at Pratt, Manhattan and Princeton, many of the top people at the firm I later established, Thornton Tomasetti, originated from my classes at these schools. It worked.

In the early days of LZA, there were about 10 people in the office. Every week on Friday at about two o'clock, Lev Zetlin would call us into his office and tell us that we had to work on Saturday and Sunday and that we needed to complete this project by Monday morning. We called it the Friday afternoon crisis. Although it was relentless, we all got used to it. At least we were paid overtime and made more money, which helped a starving graduate student. In general, whenever this happened, I adjusted my plans and did what had to be done. One time on a holiday weekend, George Feddish and his wife Louise and Pat and I made plans to go skiing at Mount Snow in southern Vermont. Of course, on Friday, Lev said that this new job needed to be worked on. I said I couldn't because I was going skiing. Lev hemmed and hawed and said he would have to put somebody else on all my projects. I said, "If that's what you need to do – go ahead." I came back on Tuesday morning and everything was fine. I learned that sooner or later I had to stand up to the boss and just say no.

On a humorous note, one day Eric Pierre, a Haitian born structural engineer, was in the toilet at LZA and noticed that there was no toilet paper. He looked down and noticed that there was a person in the stall next to him who was wearing alligator shoes. He knocked on the wall and asked for paper, but was totally ignored. He waited for the person to leave and then he hopped out of his stall into the recently vacated one. Then, Eric searched the entire office and could not find the alligator shoes. Finally, Lev came out of his office wearing the tell-tale shoes. Eric kept his mouth shut but we all remembered it.

Learning about Leadership & Management

In 1966, LZA was approached by American Airlines, Eastern Airlines, Pan American Airlines, Qantas Airlines, Trans World Airlines, and United Airlines. The 747 aircraft was scheduled to arrive and the existing maintenance hangars for Boeing 707s were too small. The facilities needed to service these much larger and longer 747s and SSTs. At this point in time, supersonic transport (SST) aircraft did not exist. We were contracted by the six airlines to provide a structural and research development project to produce conceptual designs for housing 747 and SST aircraft. This was an interesting project – for the first time, the six competing airlines actually got together to work on something jointly. What they discovered was that they were facing a major challenge not only in aircraft maintenance hangars, but also in terminal gates that could accommodate these large birds. Nothing in the United States could handle these behemoth aircraft. LZA submitted their final report on April 14, 1967 to the six airlines.

Shortly thereafter in 1968, American Airlines contacted our company and asked us to compete against Buckminster Fuller and Sadao to develop a super bay hangar design that would be a prototype to be constructed at 10 locations in the US. The lead conceptual architect was Simon Waitzman, who became a very good friend of Richard Tomasetti and me. Richard Bonner, backed up by Reed Mowry and Bob Hullett and several other American Airlines personnel, managed the project for American Airlines. By this time Lev Zetlin, Vincent DeSimone and James Chaplin had formed the company named Zetlin, DeSimone, Chaplin (ZDC). We all sat down to discuss who we would add to our team if we won the competition. Initially, it would just be ZDC and an architect of our choice. We would also add the other major disciplines, including mechanical, electrical, plumbing, fire protection, and site civil. Although we would have had an overall project manager and the lead architect, I was to be the principal person in the separate office.

I left the main LZA office and moved to 31 Union Square West. One of our favorite architectural firms was Conklin and Rossant (C&R), which was run by Bill Conklin and Jim Rossant. We decided that they were the firm to join us in the competition. We submitted our modular light gauge hyperbolic paraboloid shell roof design – and we won in a competition against Buckminster Fuller & Sadao. Most people have heard of Buckminster Fuller, the engineer philosopher who invented geodesic domes and "buckyballs" and Dymaxion cars and houses. Since he was into domes, we had the feeling they that he would submit a dome, which really makes no sense for a hangar on a tight site. We submitted our design and American Airlines picked us. I remember the evening of the day we won the competition. I went home and told my wife that I had good news and bad news. The good news was that we won the American Airlines Super Bay Competition. The bad news was that the travel required was going to be extensive. I actually made 52 round trips to California in a two-year period. That's probably why I am a 3-million-miler on American Airlines.

In 1968, I expanded the Conklin and Rossant office and started hiring for the project. This was my opportunity to start from scratch and staff up with the right kind of people to take the company into the next four decades. I got out from under the archaic management of LZA and ZDC and had enough freedom to do it the right way. This joint venture allowed me to escape the inefficient and unworkable hiring practices and calculation methods in the main office. In 1968, I hired a crack team of structural engineers, including Dr. Joseph Thelen, Joseph P. Denny, Ashvin Shah, as well as my college classmate and fraternity brother,

Richard Tomasetti, who could solve the mathematics and analysis of the hyperbolic paraboloids. Richard had followed me to NYU and gotten his master's degree and he had been as captivated as I had been with Lev Zetlin. This project gave me the benefit of running the joint venture that employed architects and engineers. I learned that it really wasn't that difficult to manage something that was multidisciplinary. It was probably one of the most creative periods of my structural engineering life. The American Airlines hangar was probably, in my opinion, my greatest elegant solution.

I was 28 when we started the American Airlines Hangar Project and 32 when we finished it. When we started, we were instructed to prepare designs for Los Angeles – LAX, San Francisco – SFO, New York – JFK, Chicago – ORD, and Dallas – DFW. So the Joint Venture of ZDC-C&R was formed. The California locations had light roof loads and strong earthquake forces. Chicago and New York locations had heavy snow loads, strong winds, and light earthquake loads. Dallas was the easiest. Our design started out based upon a scheme that could be built anywhere in the world.

At the first big meeting to meet the contractors, Niles Carr of Haas & Haney slowly came into the American Airlines facility and asked, "Is Dr. Thornton coming?" Niles was probably 20 years older than me. I responded, "I'm Dr. Thornton." He went on to say that he expected somebody with white hair and a beard.

Jim Rossant, of the architectural firm Conklin and Rossant, was not only a great architect and designer, but he was a great man. We hit it off immediately. One day it was decided that somebody had to go and study American's existing facilities on the West Coast and Jim and I were elected. Neither of the two firms had ever designed an airlines hangar before. Jim and I flew to California, arrived at the hangars at around 11 a.m., introduced ourselves, and said we would like to look at aircraft in the hangars. The first guy we met reacted in a strange way saying, "You don't understand, the birds fly during the day. If the birds are in the hangar, then

AA is not making any money. Come back at night." At least the executives who hired us weren't there to watch this display of ignorance! Having nothing to do, Jim and I took a taxi to downtown Los Angeles. Neither of us had ever been to California. In 1969, there was no downtown LA, so we ended up near Bunker Hill. Jim said he wanted to dip his toe in the Pacific Ocean, but we both had suits on. We jumped on a bus on Wilshire Boulevard and for $.62 each we took a 17-mile bus ride. We got off just short of the Santa Monica Cliffs, walked down the stairway, rolled up our pants legs and waded into the cold water of the Pacific for a short dip of our toes. We then headed south and walked until we hit Venice Beach. It was time for lunch and somehow we ended up in the Oar House, a famous watering hole for California's Hells Angels motorcycle gangs. So Jim and I are sitting at the bar trying to figure out how to get out of there. We finally called the cab and went back to the airport. The birds had arrived in the hangar, so we did our work and then we flew home. That was the beginning of a great relationship with Jim Rossant, one of the most talented design architects I have ever worked with. His sketches were awesome and his ability to listen to advice from consultants was equally as impressive.

I learned two important management lessons from the American Airlines Hangar Project. I had concerns that I would lose my spot in the LZA hierarchy when I had to leave the main office and go to 31 Union Square West to run the ZDC joint venture. Actually, the reverse became true. While I was running the joint venture, several of the other key staff came and went and paved the way for me to move into the top slot. I learned again that taking a risk and going for it, paves the way to the top. The other lesson I learned was about being thrown in the deep end of the pool – a common LZA management practice. While working on the American Airlines project, we needed a person to sign and seal the structural drawings for the American Airlines hangar. Lev Zetlin was not licensed in California and the state requires a special exam. When asked how this would be solved, Zetlin told American Airlines not to worry, that I was going out to take the test in two weeks. Talk about pressure.

The whole world knew I was going up to take the Professional Engineer (PE) license in California. If I failed, we had a big problem. I passed and became a California PE. I never got around to getting a Structural Engineer (SE) license because they wouldn't let me take the SE exam until I was PE for four years. I ultimately did become an SE in Illinois.

During the American Airlines Super Bay Hangar Project, I met Ken Hiller, one of the sharpest, most creative structural engineers, specializing in structural steel, I have ever known. He is a good friend and a delightful person to spend time with. Ken worked for a company called Bethlehem Fabricators, a steel company that wanted to secure the contract for the structural steel on the hangars, including the one planned for JFK in New York City and at O'Hare Airport in Chicago.

Ken and his wife, Earline, lived in an Edward Durell Stone contemporary home in North Stamford, CT, not far from my home in Pleasantville, NY. We often had dinner locally between 1967 and 1978. By this time Ken and Robert Oxhandler had formed the company Oxhandler Structural Enterprises. We started working together on some structured parking projects and as our relationship developed, in 1972 we formed a company called Zetlin-Oxhandler, a design/build company specializing in parking structures and other unique facilities.

This design/build company worked on several projects. After several years, however, we concluded that signing bid and performance bonds personally was not in our best interest. So we closed the company. By this point, Ken was actually president of Zetlin-Oxhandler and a full-time employee. Ken ultimately joined Lehrer McGovern and became a big fan of Thornton Tomasetti. Ken was instrumental in helping Thornton Tomasetti get many projects in the New York area.

In 1977, when Richard Tomasetti and I bought Lev Zetlin and Associates, Robert Oxhandler made an offer we couldn't refuse, saying, "If you run into cash flow problems just call me!" There was a point where we were short on cash and we called Bob and said we needed $20,000. Bob hand delivered a check for $20,000 without batting an eyelash or signing any kind of a note. We paid him back within several weeks. That's the kind of friends we needed.

In 1986, "Engineering News Record" decided to do a cover story about Richard Tomasetti and me. The article was written by Janice Tuchman. When Ken heard about the cover story, he insisted on being interviewed by Jan. Ken always had nicknames for everybody. Mine was Abu Ben Tensing, based upon the fact that I was jet setting to Saudi Arabia and had climbing experience in the Catskills. The 'Tensing' part of my nickname came from Tensing Norgay, Sir Edmund Hillary's Sherpa guide in his first ascent of Mount Everest. Unbeknownst to Ken, all the writer, Jan Tuchman, wanted was a quote using our nicknames. Jan and Ken had lunch and they probably spoke for two hours. He was really disturbed when the only line in the article was the reference to our nicknames. From that day to the present, Jan is known as Lizzie Borden, the ax murderer, for axing Ken in the article. Ken and Earline are still our friends today.

Failures

I awoke from a massive crash. Was it an airplane crash? A truck crash? A car crash? I was surrounded by debris. The debris was pieces of light gauge hyperbolic paraboloids – the structural system we had invented for the American Airlines super bay hangars. It took me a moment to figure out that it was a dream. The awesome responsibility that a structural engineer takes in general practice is relatively minor on tried-and-true, done before systems. But, inventing and patenting a brand-new system, using off-the-shelf long span cold-formed light gauge metal decking, was really pushing the state-of-the-art.

About that time, we were undertaking wind tunnel testing at Purdue University's wind tunnel facility. An early wind tunnel model of the Super Bay Hangar roof curled up like tinfoil. Professor Palmer from Perdue University was facing the other direction and was not able to detect the initiation of the failure. It was my job to inform American Airlines top brass that a $40,000 wind tunnel test failed and we did not know why. And furthermore, they would have to do a larger scale test, for even more money, if they wanted to feel secure that the design would work. They did it and we moved on and ended up constructing two hangars – one in Los Angeles (LAX) and one in San Francisco (SFO). They were masterpieces, winning every engineering award in 1972, putting LZA on the map.

In my entire engineering career, I have not experienced many engineering failures, but one occurred at Wolf Trap which really woke me up as to how the material suppliers in the United States do not tout the shortcomings of their products. In 1970, in the midst of the American Airlines Hangar Project, LZA was hired to work on the Filene Center at Wolf Trap Farm Park. This was an outdoor theater, owned by the National Park Service, to be constructed in McLean, Virginia near Washington, DC. It was a long span open-air structure of about 150 feet, patterned on the great venue in the Berkshires – Tanglewood. Approximately 5000 people would sit under its roof during the spring, summer, and fall. The architects on the project were McFadyen and Knowles. The associate in charge was Alfredo Devido, who would actually end up designing my house almost 30 years later.

One day I got a call from the Wolf Trap site informing me that as they lifted the pre-assembled 150-foot long Queen Post Trusses, an 18-foot long crack formed at the reentrant corner at the connection I developed. In order for timber to work structurally and for the bolt tables in the American Institute of Timber Construction (AITC) manuals to be valid, you have to assign some value to the allowable cross grain tension stress. The end connections on these 12-inch x 78-inch deep Douglas Fir Glulam girders relied upon transmitting the load horizontally and vertically into the wood from the structural strand cables. I searched high and low for an allowable value of cross grain tension and couldn't find one. So I backed into it by extrapolating the loads in the AITC Manual and came up with a number, albeit small. When the crack occurred, I did some quick research and found that there was a heated argument going on within code circles that the AITC should come out and say that the allowable value of cross grain tension stress in Douglas fir should be zero. I felt betrayed by an industry supposedly represented by professionals. This shattered my confidence.

As far as I was concerned, Lev Zetlin left me in the deep end of the pool on this issue. He did nothing to help me. I went to all the meetings to resolve the matter alone. I was feeling that I had convinced everybody that this was just a situation where the moisture content of the wood must have not been correct. On about the third meeting, Dr. Siev, a principal partner of Severud Associates who was consulting on the job and who was familiar to me, showed up at a meeting. I now know they realized that we had a design problem. He was a gentleman and a professional all the way through the process and we resolved the issue. We informed our insurance company of the situation. It was fixed and we moved on – even though at a couple of points during this process, I was thinking of changing careers. I felt shattered that something I had designed had failed. Fortunately, no one was injured and no one was killed. I learned that the deep end of the pool is a lonely place to be. Once I took over the management of the company, we never allowed anyone to be alone in the deep end of the pool – we jumped in and kept them afloat through whatever was going on. All firms have situations where they make mistakes, but management has to step in and help resolve the issues.

The Wolf Trap budget was $1.4 million. Approximately 10 years after its construction in the early 80s, it burned to the ground, because the National Park Service did not include fire suppression systems in the building. As you would expect with government agencies, there was "the close the barn door after the horses have all run out" approach. The reconstruction took approximately $25 to $30 million.

Christo

In early 1970, Christo Javacheff and his wife, Jean Claude, contacted Lev Zetlin and asked the firm to do some work on Christo's public art. Initially, Frank Marino handled the account. Christo had a building in lower Manhattan, which was his studio. As a typical engineer not appreciating art, Frank Marino innocently asked when seeing a pile of magazines wrapped in fabric in Christo's front hallway, "Do you collect magazines?" Of course, Christo almost died since he viewed the wrapped magazines as artwork. What a great way to start a relationship.

In 1970, Christo asked us to be the engineers on the Valley Curtain at Grand Hogback, Rifle, CO. The span was 1,250 feet. The height at the center was 185 feet and the height at each end was 365 feet at the rock anchor supports on the mountain. The actual fabric curtain was 200,000 square feet of nylon polyamide fabric and 110,000 pounds of high-strength structural strand cables.

When Frank Marino left LZA, I took over the project. This is probably my best example of why you should never work for an artist. It's also a great example of the inability to communicate with artists and the importance of putting things in writing. We designed the curtain to span between two mountains in Gap, CO. The curtain consisted of nylon fabric, four large 4-inch diameter structural strand cables connected to rock anchors embedded in the side of the mountain up high enough so that the curtain would fit in. I remember a meeting with the Colorado State Highway Department where we introduced the concept. You can imagine a state bureaucrat's reaction when we requested permission to put a curtain over a state highway with a hole to allow the highway and vehicles to pass through it. They almost died.

As usual the project fell behind schedule. Christo was collecting money to pay for it from his benefactors. By this time, winter was upon us. The contractor requested something like $80,000 to skid a drill rig up the side of the mountain in the snow to do borings. Since we had to drill the rock anchors anyway, I made the decision to wait until spring to perform the boreholes as probes and then adjust the rock anchor design length number in the spring. It seemed like a good idea at the time.

Good old Dr. Thelen visited Gap and decided not to climb all the way up to the top. I didn't blame him. So when we drove the probe holes, we found talus or loose rocks instead of solid bedrock. To accommodate this discovery, the rock anchors had to be elongated, which added money. This disturbed Christo and his wife to the point that they fired us. This was the only time in my career as an engineer that we got fired.

Our design had provided enough billow or sag so that the stresses in the fabric remained within allowable values and would not rip. After we got fired, they hired someone else who eliminated the billow and when they finally lifted the curtain, a mild 7-mph zephyr came down the canyon and destroyed the curtain. Back to the drawing board and back to the donors for another year. A documentary film directed by Ellen Giffard, Albert Maysles, and David Maysles won an Oscar. It shows the whole process and the failure. Christo also had an exhibit in Washington, DC at the National Gallery of Art a couple of years ago. I have a copy of the poster that used our anchorage drawing and didn't give us credit. At the time I wanted to sue the Christos, but why waste money on legal fees. Thank God we were gone.

Richard Tomasetti

When I entered Manhattan College in 1957, Angelo Tomasetti, Richard Tomasetti's older brother, was in my class. In those days, all classes were made up alphabetically, so everybody in my section was P to Z. This meant that the Tomasettis and Thorntons were together. Although Angelo and I got along quite well together, we really weren't great friends. He was from Brooklyn and I was from the Bronx, which meant there was a world of difference between us. In 1961, my senior year, Richard was a sophomore and pledged in my same fraternity. We became pretty close friends. Richard ultimately became president of the fraternity and showed strong leadership qualities. When he decided to go to NYU to get his master's degree, we reconnected. At that time, I had already been married for two years and I believe he was about to get married to his first wife. He moved to Groton, CT, after NYU and joined the Electric Boat Division of General Dynamics to help solve why the nuclear submarine Thresher went down with all hands on her maiden cruise.

Partnerships in both business and marriage thrive on opposites being attracted to each other. Richard and I are about as opposite as two people could be, which is probably why Thornton Tomasetti succeeded. Richard is slow, deliberate and risk averse. I am impetuous and a risk taker. I have a short attention span and am a quick learner. Our attorney, Peter De Gaetano, basically said that Richard would've been a great attorney. He could find things in contracts that even Peter couldn't find. Richard joined LZA at the beginning of the American Airlines Super Bay Hangar joint venture. In addition to working on the hangar design, he was charged by Lev to start Environ Space Research and Technology Corporation. This was part of our first strategic plan – to be a technology company, not just a commodity engineering firm. Since Richard worked at the joint venture, he reported to me. Richard and Ashvin Shah worked on developing the theoretical mathematical models for orthotropic light gauge steel hyperbolic paraboloid shells. When the joint venture was finished, Richard returned to the main office with me. Between the end of the joint venture and the acquisition of the company by Gable Industries, Richard worked on his things and I worked on my things. As Gable took over, we started working very closely together. Over time, during his five-year contract with Gable, Lev Zetlin basically lost interest in the company, collected his money, and was not seen very often in the office. He was cooking big deals in exotic places – at least that's what he said.

Gable Industries

By the early 70s, the Arab oil embargo created one of the steepest recessions in my career. There were essentially no construction projects in New York. After the country's recent recession of 2008, some famous politician said, don't waste a crisis. In 1971, we didn't waste a crisis. Lev Zetlin sold LZA to Gable Industries that year. Between 1971 and 1977, it was great to have a big daddy with a lot of money. Gable Industries was a New York Stock Exchange conglomerate controlled by Fuqua. By this point Vincent DeSimone and James Chaplin had left the firm and formed DeSimone Chaplin Engineers. In addition, for various reasons, a lot of the really key staff that had joined the firm back in 1961 had left the employment of Lev Zetlin. This opened up an opportunity for me to rise to the top of the management chain. The good news was that Lev had convinced Gable Industries to bring in management consultants to teach us how to run a company. At the time, I thought this was the biggest waste of money, but in retrospect, it was one of the smartest things they ever did and it was the beginning of the great leap forward of strategic planning that led to the creation of the largest structural engineering firm in the US. During this period, I wrote a white paper about the right way to staff and run a structural engineering firm. Although Lev Zetlin had a five-year contract, he left Gable Industries around 1974 and I became president, with Richard Tomasetti as Executive Vice President.

Office Hikes

When Richard Tomasetti and I took control of Lev Zetlin Associates in 1974, we adopted all of the good points of the Lev Zetlin approach to business, which were creativity, strong marketing, strong business development, strong promotion, and extensive publishing of our projects in journals and magazines. At the same time, we improved upon some of the weaker business practices of LZA, such as little attention to detail, poor drafting, and a lack of attention to our clients. I thought that my role from 1961 through 1966 was to be yelled at every day by our architectural clients. We refocused the entire company on quality of documentation, accuracy of calculations, and strengthened communication with our clients. We also introduced transparency in the firm, so that all members of the company felt they were part of the company. These were not elements in the Lev Zetlin regime.

When you break down the typical lives of people in an urban setting in the United States, about one third of the time, approximately 8 hours a day, they sleep. They spend another 12 hours a day working, so the balance of the time is spent with family, friends and play. Between 1961 and 1963, between being a full-time student, essentially working at least 32 hours a week and raising the family, I had spent a lot of time in the office on Saturdays. Between 1963 in 1977, the staff of LZA continued to spend a lot of time in the office on Saturdays.

In 1966, however, after I received my PhD, I started seriously hiking and climbing in the Catskill Mountains and about the same time my brother Bill and I became serious canoeists, concentrating mostly on the Adirondack rivers. The first integration of my recreational pastimes and work were the company hikes. This was a way, twice a year between 1966 in 1986, to accomplish team building exercises with the entire staff of LZA and later, Thornton Tomasetti. What better way to get to know your staff than to spend Friday evening through Sunday at noon having fun climbing the Catskill Mountains, playing flag football, eating and drinking? After the company grew to a size too big to invite everyone, we switched to the Master's Hike—where you had to be over forty-years-old to participate. We then moved to the Berkshires and used the Jiminy Peak Condos we had purchased with friends and relatives as home base for the hikes.

Driving up from the New York State Thruway on Route 32 revealed the Northern Escarpment of the Catskill Mountains. These mountains were painted by Thomas Cole and the Hudson River School landscape painters. Unlike the Adirondacks, the Appalachians, and the Rockies, which are all formed by plate tectonics, the Catskills are actually not mountains. They are in an eroded ocean bed. As the glacier moved through the area, it gouged out the soft stuff and left the hard stuff. The Catskills have no synclines and anticlines, which are folds in the rocks. The stratification is horizontal, and on close inspection, one can see fossils embedded in sedimentary rock. The Hudson River Valley is at an elevation of about 600 feet. The tallest peaks in the Northern Escarpment are 4,000 feet.

From 1956, when we first bought the Catskill land, until 1966 when I began to hike, I always looked up at these mountains and said, "I guess I better climb them because they are there." So after a weekend in the Catskills with my family, I went into the office and I said to my colleague George Feddish, "Do you want to climb the Catskills with me?" He said, "Of course, I can wear my Austrian hiking boots." We hiked twice a year – the last weekend in April and the last weekend in October. We picked these dates because there were no bugs and the weather is subject to very interesting, potentially variable conditions, which made the event much more exciting. So in April 1966, my brother-in-law Ray Podaski, George Feddish, (Manhattan '58), Artie Ludwig (Manhattan '58), Gunther Arndt (a structural engineer), and I climbed the mountains. We were backed up by Jack Grad, a draftsman, and our cook, and Arpad Tarkoy, our chauffeur and the son of a Hungarian general and militarist who fought in the German army on the Eastern front. Arpad and I were both under the Aries sign. He told me that all the great men of the world were born under the sign of Aries – Hitler, Khrushchev, Arpad, and me. He was serious. When my first child Diana was born and I was handing out cigars in the office, Arpad said to me "Girls are junk." I remembered my father saying, "There is no such thing as a bad woman. Bad women are the result of bad men." Eventually having three daughters, I know that girls are far from junk.

In preparation of our hike, we did our homework. We got the maps and trail maps and we evaluated the time it would take to climb from elevations of 600 to 4,000 feet. The names of the mountains on the North Escarpment were Thomas Cole, Black Head, Black Dome, Acra, Burnt Knob, and Windham High Peak. We started out in Acra, NY, with light clothing and we discovered that all of our calculations were wrong. When we got to 1,900 feet there was a blizzard. We were all soaked and potential victims of hypothermia. We found a cave, started a fire, dried out, and ate lunch. Our lunch was Pepperidge Farm soups on Sterno stoves and cognac. The sun came out and the snow was hanging on the hemlocks. It was like we were in the High Sierras. It was absolutely gorgeous. We did make it to a notch at 1,900 feet and dropped down into Batavia Kill. We did a forced march down the road to meet Arpad, who of course knew we were coming but didn't drive to meet us. He was waiting for us in Hensonville in the car. Jack cooked us an unbelievable dinner and we drove back to New York City.

A typical weekend consisted of an international food festival on Friday night at the house with Chinese, Thai, and Greek food. Dinner would begin upon the first arrival and would go to the wee hours of the morning. We awoke at six a.m. and had breakfast and then drove to the mountain to hike. Premala Raj, a brilliant Cornell structural engineer with the company, became the first woman to break the sex barrier and join the hike. Walking up the mountain with Premala, I asked her about arranged marriages in India. She said she didn't do it, and instead married out of her sect. I told her I thought there was a low divorce rate, and she replied that it didn't mean couples were happy.

The weekend also included touch football with blocking on Saturday afternoon and Sunday morning. During one touch football game, I threw Richard Tomasetti a game-winning touchdown pass, which he caught by jumping up and coming down on the ball. He broke two ribs.

On Saturday evening of the weekend, we would drive into town to either Coxsackie, Cairo, East Durham, or Wyndham for dinner. After dinner everyone would gather in the living room, family room and kitchen to play cards and drink. Bob Liss joined the company around 1967 and became a consistent attendee, along with Jim Chaplin and Vince DeSimone. We really got to know our employees on these hikes. It was a great way for new and existing employees to meet one another. It was very democratic and there was no hierarchy. That makes a company a company.

One memorable dinner was when 12 of us took a trip to Coxsackie and left Len Joseph behind at the house by accident. Joe Denny and I drove back to the house to go get him. After about an hour and a half had elapsed, we found Len sitting in the dining room, a little pissed off. Len had been doing the dishes and went upstairs to take a shower when we were loading up the cars. He ran downstairs as we all drove away. No good deed goes unpunished. After that year, the expression, "Where's Len?" became popular on hiking weekends. On another hike, after more women started coming on the hikes, the women all took my parent's master bedroom, which had a very large bed and fair amount of floor space. As a joke, Tom Scarangello plugged in a canister vacuum cleaner and placed it under the bed with the plug out in the living room. After Sally Handley and her crew went to bed early, Tom plugged in the vacuum cleaner for about 5 minutes. They all scurried around trying to figure out what was making the noise, but before they could figure it out, Tom would pull the plug. Tom repeated the prank. It took them a while to figure out what was going on. We all had a very good laugh.

On one hike my father actually joined us. Although he ended up a little stiff in the process, he totally enjoyed being there. My brother Bill came down from Potsdam, NY, with one of his colleagues, Ted Totten at Cives Steel and spent the weekend with us. On one of the early hikes, my younger brother Robert came. Robert was playing offense and Sam Weinberg, an Israeli who had never played football before, was playing defense in his corduroy suit jacket. Sam thought that you blocked with your arms at your side and that you led with your chest. Robert blocked Sam and broke two of his ribs. We continued the hikes until my father passed away and we sold the house.

One of the Lev Zetlin employees who always hiked with us was Robert Liss. Robert was a truly unique character who had joined Lev Zetlin Associates after coming out of the Army. He spent time at Turner Construction on Belmont Park in Queens and joined the firm in about

1965. He was my soul mate. I would ask him to go hiking and sailing and he would always say yes. One day I asked him, "Do you want to go buy a condominium at Jiminy Peak in Massachusetts?" Robert said yes. Robert and his wife, Rochelle, had a unique and strong relationship. Robert was the original free spirit who did all the things you wanted to do, including fly-fishing and bone fishing with his buddies. Rochelle was there to encourage him to be all that he could be.

We shared the condominium at Jiminy Peak and skied with Robert and his son Michael. Robert later came on every Male Bonding sailing trip and on every hike on the high peaks in the Catskills and in the Berkshires. He was a connoisseur of great white wines, a fabulous tennis player, and an avid cyclist. Sadly, Robert passed away in 2012 of an aneurysm, which occurred after a 14-day bike trip in Vietnam.

We staged these hikes from the four condominiums owned by myself, Robert Liss, Dan Cuoco, Richard Tomasetti, and Jay Prasad. We generally climbed Mount Greylock, the tallest mountain in Massachusetts. Since we were over age 40, the touch football ended and we generally had dinner in Hancock Village and all left the next morning after breakfast. We reached a point where age caught up with us and we eventually sold the condos. I then moved to Maryland and the semiannual hike was history. This was a great opportunity to be with my good friends, many of whom had sailed with me on my Male Bonding sailing trips. I continued to get calls once the hiking and sailing ended.

Washington, DC Office

During the early Gable years in 1971, Lev Zetlin and I convinced Gable Industries to bring in construction management and project management, headed by our friend Richard Crowley of Total Project Management (TPM) Constructors. They were part of another conglomerate that went under, so because they were out of work, we decided to bring them in so we could become the "Bechtel of the East."

Right after the acquisition of LZA by Gable, we decided to open a Washington, DC office. A great man named Brigadier General Charles Cogswell, who was retired from the Marine Corps, was selected to be in Washington, DC. I, along with Richard Tomasetti and Joe Zuliani, started going to Washington and spending a lot of time there. I attribute my selling ability and business development skills firstly to Lev Zetlin and second to Charles Cogswell.

Charles was a highly decorated Marine general who was in the first landing on Guadalcanal, one of the bloodiest battles of World War II. He also served on the staff of Admiral Chester A. Nimitz. While serving in the Pacific Theater, he earned the Silver Star, the Bronze Star, and the Purple Heart. Charles was a gentleman and a soldier. I became very friendly with Charles and went fishing on the Chesapeake Bay with him in the early 70s, along with my young son Charles, and our lawyer, Peter De Gaetano. We didn't catch a lot of fish, but we had a good time.

Charles Cogswell taught me and my colleagues that it was simple to win US government design contracts – you just had to be there. To do so, we spent a lot of time at the Army and Navy Club. We also attended many industry dinners and visited numerous government agencies. This was the period of the Arab oil embargo, so it was the Washington and federal government work that kept us going strong through the 70s.

Charles introduced us to the Order of the Carabao and the Wallow of the Carabao – yes, just like the Filipino water buffalo. The Order of the Carabao was a retired military organization exclusively for veterans of the Pacific Theater wars. The party, called the Wallow of the Carabao, included such members as Admiral Bull Halsey and Admiral Chet Nimitz, who were still alive at the time. This black-tie dinner, in the largest ballroom in DC, rivaled the annual Gridiron Dinner where the Washington Press Corps would roast politicians. When the best of the Marine Corps drum and bugle corps marched in and played, the hair on the back of my head stood up. Cogswell ultimately became the Grand Paramount Carabao, the boss, whose main responsibility was watering the herd. These dinners went on for about 10 years. I really enjoyed the tradition and pomp and circumstance of the military, as well as the opportunity for networking. There were several prominent politicians in attendance as well.

My first engineering mentor was Eugene Fullum whom I met during one of my summer jobs. He was a young Marine about seventeen-years-old on a troopship about to invade Japan in 1945. When the US dropped the bomb, Gene returned to the US. I started inviting Gene to these dinners and he loved them. We would meet at a LaGuardia Airport, fly to Washington DC, check into the hotel and attend this wonderful dinner with Charles Cogswell and other LZA staff members.

After Charles retired, he remained in Clifton, VA. He passed away in 1993. I attended the full burial service at Arlington National Cemetery. This was the one and only time I was honored to be able to attend such a funeral – it was awesome. Charles was a humble man who never talked much about what happened on Guadalcanal. When he was honored at one of the dinners, I sat there in amazement and awe as they announced the accomplishments that this gentleman and soldier attained. Charles was a great man.

As the Gable year commenced in 1971, it was a very interesting experience for Richard Tomasetti and me. We made most of our sales targets and we hit our profit numbers. We worked with Case and Company, a management-consulting firm from Cleveland, and charted our first serious five-year plan. We learned to do financial pro formas, assets and liabilities sheets, market segmentation studies, and all the things that we needed to do to run a company. At the time I thought that the $100,000 for management consultants was a waste of time, but in hindsight, it's probably the reason why Thornton Tomasetti is the best managed and most profitable engineering firm in United States.

During this time, Lev convinced Gable Industries to start Total Product Management (TPM). TPM seemed to be looked upon favorably by Gable because Gable didn't lose more than what TPM projected they would lose. Richard Crowley of TPM was a very nice guy. I later spent a lot of time with him at Stamford Yacht Club in Stamford, CT, in the 80s.

Richard Crowley, coming out of a larger company, had a little more maturity when it came to dealing with corporate bureaucracies than we had. So eventually, TPM made an end run on us to try and gain control of LZA. We counter attacked and arranged for a meeting in Atlanta with Gable Industries and TPM. I am not exactly sure what happened, but Crowley missed his flight. By the time he showed up three hours later, I had actually closed the deal. We said that we would not welcome being managed by TPM and if that was their decision to do that, we would all quit. Not having Richard Crowley in the room gave me a great opportunity to be a lot more aggressive. We had won.

In 1974, Lev Zetlin departed. He walked in one day and said to me, "Charlie, I'm leaving." I was very surprised. Because I was number two and Richard was number three in the management hierarchy, I became president and Richard became executive vice president.

Sometime around 1976, Gable Industries questioned why a $300 million company owned a small engineering firm with $3 million a year in revenue. The two top guys at Gable Industries, Duke Waddell and Neil Schactel, came to me and said they wanted to sell us the company. Even though we told them we didn't have any money, they said not to worry and that they would make it work. I turned to Richard Tomasetti and said, "Let's do a 50/50 ownership." So we did a leveraged buyout, paying approximately $500,000 for the company – no money down, no personal guarantees, and a five-year payout. Richard Tomasetti, Peter De Gaetano and I did battle with Cravath Swain and Moore, a prominent New York law firm that probably billed four times what Peter charged us, and we got the company. As long as we owed money to Gable, we were not free to do anything. Everything was subordinated under our debt to Gable. We paid off the debt in two and half years and the rest is history.

Outward Bound

In September, 1977, John Raynolds, CFO of Heede, was the jacking contractor on all of David Termohlen's International Environmental Dynamics (IED) buildings. John was on the National Board of Outward Bound and invited me on an Invitational Outward Bound trip. Outward Bound had many locations and they had different offerings, ranging from 3-day short weekends to 7-day trips to 21-day trips. An invitational was a typical 7-day Outward Bound trip in which they invited potential donors to the organization. Because of this, my only cost was my airfare to get to Moab, UT. There were 30 people on the trip, with the youngest being about thirty-years-old. I was thirty-seven-years-old and one guy was actually close to 80. There were 22 men and 8 women. We spent 7 days in Cataract Canyon and on the Colorado River. We put in our rafts at Arches National Park near Moab and took out at Hite Marina on Lake Powell.

We first met in Grand Junction, CO, for dinner to get to know one another. We woke up the next morning and jumped in vans and were driven to Arches National Park. For our trip down the Colorado River, five guests were assigned to each inflatable raft with a guide. The guides explained that they were there to protect us from killing ourselves, but that navigating the rapids, which we would encounter on days two, three, four and five, was our responsibility.

There was a big raft with the motor on it that was piloted by Huck, a muscular hunk of a guy, right out of central casting. He rowed and motored the raft with all the food. All of the guides were skilled and attractive and made it all look so easy. Cigarettes and booze were outlawed. Everything we took in, we carried out. This meant that the Porta-Potty we used for seven days had to be taken off the raft on the last day.

It was a real experience, because people in general are fairly modest. On an Outward Bound trip, modesty disappears on the second day. On the first night, we put the Porta-Potty up behind the woods to protect our privacy but by the second day, it was in plain sight. The first night I laid down my sleeping bag on the riverbank near some driftwood and a scorpion came out of the driftwood. Not wanting to be stung by the scorpion, I killed it. The next morning they told us not to kill scorpions, as they really were not that dangerous. I never told anybody I killed one. It was September and there was no humidity and no dew, so we slept under the stars, which was unbelievable. If I had been camping in the East, I would have woken up in the morning soaking wet.

On the first day of the trip we learned how to ferry the raft – how to manipulate the raft to go straight, backwards and sideways, and right or left. The next morning we woke up and we hit the first day of heavy rapids. I made the acquaintance of Susie Baker, a teacher from Greenwich High School. Susie was an American Athletic Union (AAU) diving coach and actually almost equivalent of an Olympic diver. So we appointed Susie Baker as the captain to go through the first rapid.

Our guide stopped the raft above the rapid and told us that as a team we needed to decide where we would enter the rapid and how we would traverse it and come out of it at the other end. We went through the first rapid backwards with Susie losing her self-confidence and losing control of her crew. By the time we hit the second rapid, Susie was beside herself and didn't want to be captain again. We said, "Susie, you are the captain," and she took us through the second rapid perfectly. We all took our turns captaining the raft through the rapids. The river was not flowing at full flow, which in a way made the trip easier; although there were more rocks exposed that we had to dodge.

Each night, we pulled the boats up on the beach and had dinner. The food was prepared by the guides and was spectacular – this was less than roughing it. One night, we built a big fire inside a pile of rocks. Afterwards, we took the raft paddles and constructed a tent over the rocks and poured river water on the heated rocks to make the steam. We had our own sauna. Some of the people also took a mud bath. I have a picture of Susie in her bikini, wallowing in the mud. I didn't do it because I had enough sand and dirt in my shoes and socks already from swimming in the river. Wallowing in the mud was not my idea of improving the situation.

On the second night of the trip, Leo, the president of the Ford Brothers Foundation, came to the campfire and fessed up he had smuggled a carton of cigarettes on the trip. He admitted his guilt and threw the cigarettes in the campfire and never smoked again. He told us the only reason he was there was that his daughters had all done the trip and they had set him up to be on the trip a couple times, but he backed out. He did not want his daughters to say he backed out again and was a wimp.

After about four days of rapid running, we climbed 3,000 feet up the canyon wall with 40-pound packs. This was apparently the same trail that Butch Cassidy and the Sundance Kid came down on horseback to evade the Federales. Although years before, after my 40-mile hike with the Boy Scouts, I had sworn never to backpack again – we did make it to the top. Next, we climbed an 80-foot sheer rock face with harnesses and repelling devices. The guides said we would do it twice and the second time would be a surprise. The first ascent, I was tense, nervous, and every muscle in my body ached, but I made it to the top. Everyone did it, except one person who refused to do it.

Then they told us that the surprise was that on our second climb, we would be blindfolded. It was amazing – there was no tension, no sweating, and no pain. Why? We were now in the hands of our teammates. It was amazing to see the effect of relying on your team. This is what life and work and running a company is all about. Without a team, you will never succeed. This is the principle upon which the ACE Mentor Program would also be founded.

They then took us up into the Doll House area of Pinnacles in Canyonlands National Park for each of us to do a 24-hour solo. They gave us each a sleeping bag, a canteen of water, and an orange. You could take a camera, a notepad, and a pencil. I cheated and took a roll of toilet paper too because I wasn't going to use cactus. So they dropped us off about a quarter-mile apart from one another. It was September, so it got dark quite early. I had noticed before it got dark that there were cougar tracks all over the place. They told us not to worry about cougars, as they were more afraid of us than we were of them. There wasn't a cloud in the sky that night and the atmosphere was totally devoid of humidity. All night long, there was a stunning array of shooting stars. I finally went to sleep and woke up with the sunrise. At first light, I took pictures of the Sangre de Cristo Mountains and waxed poetic in my head about what it was like to be alone for 24 hours. Most people have never been really alone for 24 hours. Thank God it wasn't the three-day solo Outward Bound used to do, where you had to forage off the land eating berries in the high desert.

They picked us up and the first person I saw was Susie. I asked her what she did for 24 hours. She said she spent the entire time naked. I told her she should have called me and she laughed. The next person we picked up was the doctor, an Outward Bound regular. He told us he had something to show us. He took us behind a big rock and there was a sidewinder rattlesnake coiled and ready to strike. I asked him how he found it and he said he went behind the rock to take a pee. I asked him why he did that since there was nobody around. I guess we are all creatures of habit.

We gathered the rest of the group and headed to the highlight of the trip – rappelling off a 150-foot cliff – that's the size of a 15-story building. I had never done this before. My guide, whom I called "Heather the feather," didn't weigh more than about 100 pounds. As I backed over the edge with the rappelling lines, doing it the way they had showed us, I told Heather I was frozen and wanted to come back. She told me I was past the point of no return and had to go. I said okay and I slid down 150 feet. I wanted to go again, but time ran out. We got back in our rafts and spent the rest of the trip motoring day and night to get down to Lake Powell.

Every year at the Plaza Hotel in New York City, Outward Bound had its black-tie fundraiser. The men wore tuxedos with hiking boots and the women wore beautiful gowns with hiking boots. Guests of the event could rappel off the grand staircase in the grand ballroom of the Plaza Hotel. It was really a gas. I went for about three or four years and then sort of lost interest – but over the years, they got their money back for taking us on Outward Bound.

If I had to attribute one single event in creating my can-do attitude towards life – Outward Bound was the event. Although I was a self-confident kind of person and a risk taker, Outward Bound really propelled me into believing that I could do anything. I came back from this trip and there wasn't anything I could not do.

My son Charlie, as well as three of my grandchildren, Casey, Kaitlin, and Brandon, have done Outward Bound trips. I really recommend to anybody that can afford to do so, to take an Outward Bound trip, especially down the Colorado River.

When I got back from my Outward Bound trip, I was having lunch with Will Turner, who was about sixty-years-old. He was a super salesman for the paint company Tnemec, which is cement spelled backwards. He was a good friend and he told me that after 35 years of marriage, he was getting divorced. He was not happy about it. I told him about Outward Bound, so he went to the Florida Everglades and did the trip there. In the same pulling boats as Will was in, there were three or four young women. They slept on the oars and went to the bathroom over the side of the boat in front of everybody. He said it really took about a day for his modesty to go away. Unfortunately, he had a detached retina during the trip and had to be airlifted out by helicopter. When I saw him later, he was having dinner that evening with the girls from Outward Bound and seemed to be on cloud nine. He is now a pretty happy guy. Outward Bound changed his life.

Hartford Coliseum

It was 6 a.m. on the morning of January 18, 1978 and I was driving from my home in Pleasantville, NY to the White Plains train station.

On the radio an announcer stated that the Hartford Coliseum in Connecticut had collapsed. It couldn't be! Entire 300 by 360-foot buildings do not collapse. Usually, a shed, canopy, porte cochere, or an adjacent ancillary building collapses due to snow shedding and drift, but not the main building. When I arrived in the city, I walked from Grand Central Station to the office and was told by several people that Hartford Coliseum's entire roof had collapsed that morning. I couldn't believe it.

Sometime that morning our receptionist informed us that the City of Hartford was calling us. One of my partners, Donald Griff, took the call. The City Manager wanted to schedule an interview to hire an investigating engineering firm the next morning, Thursday, January 19. Several of us at the office caucused, but everybody had commitments they couldn't break the next day, so I went alone. I drove to Hartford and was met by the city manager and his team. I made a one-hour presentation about our firm's qualifications and experience to be the investigating engineering firm for the project.

They thanked me and said we would be hearing shortly. I was then given a tour of the collapsed structure by the Police Chief and the Fire Chief of the City of Hartford. It was obviously still in a dangerous condition and we had to be quite careful picking our way through the project and debris. Each chief showed me the location of their season's ticket seats to the Hartford Whalers games. In both cases, the seats were destroyed. Both men were traumatized. At the time I didn't realize it, but had the snowstorm started five hours earlier, the roof would have collapsed during a University of Connecticut basketball game and 7,000 people would have been killed. Engineering is a serious business, very serious business.

I received a call from James Dakin, Hartford City Manager at about 11 p.m. that night, asking our firm to mobilize, as we had been selected as the investigating engineering firm. He asked me if we could be there the next day.

The big blizzard of 1978 occurred on a Friday. My parents were with us on one of their trips. My father, being a retired chief inspector of the New York City Building Department, volunteered to go with me to Hartford. It was a busman's holiday and he had a ball. I mobilized several key staff members, including Paul Lew and Paul Gossen and arranged to meet them in Hartford. In light of the blizzard, the State of Connecticut was shut down. It took me approximately six hours to drive from Pleasantville, NY, to Hartford, CT – a drive that normally took one and a half hours. We had to follow National Guard trucks and plows into Connecticut.

When I arrived, the City Council and the city manager were in session. They introduced me as the led person handling the investigation of the causation of the collapse of the Hartford Coliseum. We stayed at the Sheraton Hartford and the rest of the team showed up Saturday morning. Paul Lew and I lead the investigation back in the office in New York City and Paul Gossen handled management at site. Since the main members in the roof structure were not in the "American Institute of Steel Construction Handbook," we had to recalculate the member properties, which took us a day. By Saturday afternoon we knew what had caused the collapse. The structural engineer on the project assumed that the unsupported length of the major compression members was 15 feet, when in fact it was 30 feet. This meant from a simplistic point of view that the structure was overstressed by a factor of four. Normal set safety factors in structures are more on the order of three.

The interesting thing about this assignment was that our work was to be essentially transparent. "The Hartford Courant," who more than anybody else in network TV wanted to talk to me, was told I would be totally available. A very sharp reporter for "The Hartford Courant" started to recognize my patterns on my weekly visits to Hartford and he would be waiting around the corner of the nearby library to City Hall and he would step out of the shadows and say, "What are you going to tell us today, Dr. Thornton?"

I had the pleasure of hiring my thesis advisor Charles Birnstiel to work with Paul Lew and our in-house group to do an extremely complex nonlinear, inelastic buckling analysis of the entire roof structure. The analytical horsepower that we had in-house, coupled with Charles Birnstiel and the Cold Regions Engineering Lab from Hanover, NH, a Corps of Engineers' facility, provided the technical expertise we needed. Charlie Birnstiel is one of the most intense hardworking people I have ever known. When he gave me a fee to consult with us, I doubled it because I knew he would never be able to do it for the number of hours he quoted me.

The first thing we had to do was to prove that the weight of the snow on the morning of January 18, 1978 was, in fact, the heaviest snow load ever experienced by the facility, which was about four-years-old. The Cold Regions Engineering Lab took care of this analysis for us, took roof samples, and developed a time history of rain and snow, and boring. They determined that the snow load on the roof at the time of collapse was 16 pounds per square foot of snow. It was supposed to be designed for 40 pounds per square foot of snow.

Mayor George Athanson, a mischievous character, was a little bit miffed by the fact that he didn't get to select the engineering consultants to perform the investigation. So, he went to the University of Connecticut and formed his own team of University professors. Professor Kardestunker led the team, along with several other professors whom I got to know, but Kardestunker was sour grapes because we were selected to lead the investigation.

Whenever I presented interim reports to the City Manager and City Council, the Mayor would sit across the table from me and his professors would sit right behind him. After I was finished, a professor would hand him a piece of paper, he would put it on the desk in front of them and he would read this rather technical, involved question – the subject of which he had no understanding. He would smile at me, the professor behind them would smile at me, and I would proceed to just negate the relevance of the whole issue he was talking about. This became fun. The rest of the City Council was pretty reasonable. Barbara Kennelly, one of the Council members, actually became a member of Congress after a couple of years.

We completed the final report in June 1978 and we nailed it! The roof collapsed at 16 pounds per square foot of snow and our model showed that it should collapse at 16 pounds per square foot of snow. I presented the results in Hartford in front of a live TV network that was being transmitted all over New England and maybe even nationally. This is when I learned to never, ever speak to the mainstream media – especially "The New York Times." One of their reporters came up to me after a grueling hour or two in front of the cameras and said, "I don't have time to read the report, can you tell me who shot John?" I answered, "No, read the report." She finally asked me an innocent question, "Who were the architects with the engineers, since it was a design error?" She wanted to know. After refusing to tell her three times, I named the architects and structural engineers. On the front page of "The New York Times" the next day, she quoted me as blaming it on the architects and engineers. I learned if you are ever in a situation like this, be cautious when talking to the media, as they may not be interested in the truth.

Overall, however, the publicity was exactly what Richard and I were looking for, to put our faces on the map and launch the beginnings of Thornton Tomasetti. Our country was coming out of the Arab Oil Embargo recession. In addition to "The New York Times" cover story, I was featured in multiple articles in every engineering magazine. This truly launched Thornton Tomasetti's forensic and performance services. By summer 1978, the final report was finished. It was another elegant solution.

In December 1978, I had dinner with James Dakin and his wife. I asked Jim what made him pick our firm in the interview right after the collapse of the Harford Coliseum. He responded with one of the best compliments I've ever received in my entire career, saying, "You were the only one of everyone interviewed who I thought the people of the City of Hartford would believe."

Client Sailing

So after my great success at sailing around Shelter Island in 1979, I decided to start client sailing. I would charter a big boat for a week, most probably in Westport, CT, and we would invite clients for sailing seven days in a row. We chartered Viking boats – a CSY 37-foot, heavy boat, developed for the charter trade, sailing in strong winds in the Caribbean. I picked the week, made the deal with the owner, and went into Richard's office to tell him that we were starting client sailing. Mr. Deliberate, as I called him, asked, "If we go sailing, what happens if we sink? What happens if we roll over? What happens if we lose a client?" I told him not to worry about it because I knew what I was doing and it was a heavy boat. I told him we wouldn't have any problems.

Richard was flabbergasted that I would even consider taking clients on a 37-foot boat with my limited experience. I guess I am impetuous, but I am also confident that I know what I'm doing and that I am not going to do something stupid. On the third day of the first week of sailing, the wind was blowing over 30 knots out of the North. The clients on the boat were Cushman and Wakefield, including Tom Imperatore and his wife Florence, George Feddish and his wife Louise, Mike McCambridge and his spouse, and Richard and I. Since we were sailing out of Westport, CT, and the wind was blowing at 30 knots out of the North, Long Island Sound was dead calm. If we had been on the North Shore of Long Island, we would not have gone out. Tom Imperatore was a graduate of the Merchant Marine Academy. We were bombing along at probably 6 to 7 knots with all the sails sheeted in tight. We were heeling about the way that a CSY 37 should and Tom taps me on the shoulder and says to me, "Let the fucking sails out!!!!!" I asked why and he told me to look at Florence's knuckles, which were white. She never sailed with us again over the next 10 years, but would join us for dinner afterwards. Essentially, that was my first lesson that if you depower the sails, you still go fast and you don't heel the boat quite so much. I was ready to learn.

We organized client sailing on chartered boats from 1979 to 1996. In 1996, when I bought the Tartan 4100, Elegant Solution III, it was big enough to take out 12 clients, so we started sailing my own boat. About 3 years into the client sailing at Westport, CT, Richard Tomasetti insisted that he back the boat out of the slip. Docking and undocking is the trickiest part of sailing a single-engine single screw keelboat. I cautioned him, but he wanted to show he could do it. It was Saturday morning and all the people at Cedar Point Yacht Club were on their boats. As we backed out, the wind grabbed the bow and spun Richard in the wrong direction. He was now heading toward the wrong end of the fairway, which is the space between two rows of slips. He refused to take any assistance from me and he proceeded to make a 360-degree turn in small increments like a broken U-turn. Other than touching a pile, he did pretty well and as we headed down the fairway, everybody on the boats cheered and clapped. With a very loud voice, Richard said, "Just practicing!" Richard has a great sense of humor and is a great joke teller.

5
Meeting Carolyn

"A faint heart never wins the fair maid."
—CHARLES THORNTON, SR.

I WAS SITTING IN MY office at about 5 p.m. on Friday, July 20, 1979, planning to go home and have dinner with the family and leave Saturday morning for the family's home in the Catskills. At about that time, Richard Tomasetti walked in and said, "Do you want to go bar hopping?" I told him I was getting up early the next day and said I wasn't interested. He persisted and I finally said yes.

We went to a couple places in Midtown. We ended up in the Library Discotheque in the Barbizon Plaza Hotel on Central Park South and Seventh Avenue. At about 7 p.m., I was leaning against a mirrored column, wondering why I was there, and this lovely lady came up to me and asked me to hold her glass of wine. I agreed and stood there for what seemed like an in ordinate amount of time, wondering if this woman was just leaving and playing some sort of game. It was a discotheque and you had to have a stamp on your hands to go to the ladies room, which was outside the disco and in the basement of the hotel. Evidently, she got a little disoriented and it took her a while to find a way back in to the discotheque. I would learn later that she tends to be directionally dyslexic.

She introduced herself as Carolyn Heldman and we went through the usual singles game, asking about each of our marital statuses. I told her I was a widower with three children. She thought that was great, because she was a career woman and did not plan on having children. Her current job was doing marketing research for Exxon Solar Thermal Systems in Florham Park, NJ. She had a bachelor's degree from UCLA and an MBA from UCLA and had come to New York to work for Doubleday Publishing Company, had moved to International Paper Company, and then eventually to Exxon.

As we began to get to know one another, I learned that she had just broken up with some guy who was six foot, three inches tall and who had blondish hair and a PhD – so I fit the bill perfectly. She was 29 and I was 39. Later on, she told me that she would not have been interested in me if I had been age 40. We found ourselves a very comfortable spot in the library on big soft chairs and we talked until 11 p.m. when I had to leave to catch the last train out of Grand Central Station at midnight. We walked together to Grand Central Station. As we passed several beautiful buildings on the way, including Lever House and Seagrams, I asked her if she liked the buildings. She asked me what I meant by "like the building." I realized people in the business of construction like buildings, while other people just look at them.

We stood in the grand room at Grand Central Station and chatted for a bit. I asked her if she would like to have dinner with me on Monday night. She said yes and gave me her phone number. I kissed her under the Kodak Coloramas and caught my train.

The next day, I got up early and took the kids to the Catskills for the weekend. We got home early on Sunday and the second I got home, I called Carolyn and said, "Why wait until Monday – how about dinner tonight?" She agreed. I remembered my father's expression that a faint heart never wins the fair maid. I had concluded in the time I had spent with Carolyn that she was not only attractive, but she was extremely intelligent, ethical, rational, well-educated, as well as a good conversationalist. I think I knew then that she could potentially be a great partner for life. The fact that she had never been married before made life much simpler, from my point of view. She was interested in learning how to ski, learning how to sail, and learning how to play better tennis – all things that I was interested in at that point in my life. I felt reborn.

Carolyn lived on the seventh floor of an apartment building on the southwest corner of 27th Street and Third Avenue in New York City. Our first real date was on Sunday night, after meeting her on Friday night. We had dinner at a restaurant on Lexington Avenue and 28th Street. We started dating and that dating led to more than dating. One of my first dates with Carolyn was to sail on a small 18-foot sailboat that belonged to a friend of mine. She thoroughly enjoyed it. The boat was docked right at the marina where we would ultimately buy our first home together in 1981.

My standard line to my mother-in-law, Anna, when I was with Carolyn, was that I was working late and I was staying at Richard's apartment. Both my mother and my first mother-in-law had funny senses of humor. After staying many times at Richard's place, my mother-in-law said to me, "Charles, your mother's concerned about the fact that you are spending so much time with Richard. She thinks you are gay." I thought this was very funny.

One night in the fall of 1979, Carolyn and I had dinner in a restaurant on Lexington Avenue and 28th Street. I learned a lesson that cool and rainy night. We had finished dinner and were both tired. I got my tan raincoat and we went back to her studio apartment and went to sleep. The phone rang and it was my mother-in-law. She said she had called Richard first and he said I was out to get coffee. My mother-in-law had a call from American Express telling her that I had someone else's raincoat, with their car keys in the pocket. I couldn't believe that American Express gave out my personal information to this restaurant. They then called my mother-in-law and said they needed to get in touch with Charles H. Thornton, who had just had dinner there. They explained that I had another customer's raincoat and car keys. Can you imagine if I was married and having an affair and American Express gave out my personal information? So I jumped out of bed, got dressed, and walked to the restaurant and exchanged coats. Never leave the keys to your car in your tan London Fog trench coat when you go out.

In November of 1979, the family had Thanksgiving dinner with Mike Bellanca and Janice Giles Bellanca at 324 Lake Street. Carolyn came to meet my father and mother, my mother-in-law and my three children. I guess this was intimidating to a 29-year-old woman who had never married. She said they stared at her all night. I was delighted that our relationship was finally out of the closet.

Carolyn started joining my family and me skiing at Jiminy Peak and on sailing trips the next two summers. We went to Steamboat Springs for the Christmas of 1980 and spent time with Ray Podaski and visited Uncle Bill and Aunt Agnes in Salida, CO, where we had a great time skiing Western style.

Carolyn and I became engaged in 1980. Neither of us wanted a large ceremony since Carolyn's entire family was in California and we didn't want to have them travel to us or for us to have to travel to them. So I chartered a Pearson 35-foot boat from Greenport, Long Island, for the weekend of June 6, 1981 when we would be married. We informed Diana, Kathy, and Charlie that we were all going sailing from Greenport to Essex, CT, but didn't tell them we were getting married. We didn't want them to have to keep the secret from our families. We made arrangements to stay overnight at the Essex Marina, adjacent to the Griswold Inn, the oldest continuously operating inn in America. Several weeks earlier, we had driven to Connecticut to arrange for a female Justice of the Peace to marry us on the dock at the marina at 7 p.m. We got our blood tests and our marriage license. We allowed Diana to bring a friend who lived in Old Saybrook, CT. She probably would not have invited her friend if she knew we were going to tie the knot. The kids nicknamed Diana, "Leader of the Laundromat," because she always had a plan for socializing with friends.

The day before we got married, Carolyn and I closed on the three-bedroom waterfront condominium at Schooner Cove in Stamford, CT. She asked me what would happen if we didn't get married. I answered that we would be partners in a real estate venture. Carolyn continues to credit me in making the whole thing work. I do know how to compromise.

So, we all slept on the boat that night and sailed back to Greenport. Carolyn called her mother in California that night and I called my mother the next morning to announce that we were married.

After Carolyn and I were married, I left my mother and father and first mother-in-law in the house in Armonk, so that Diana, Kathy, and Charlie, when they came home from their respective schools, would still have a home. Carolyn and I moved into the waterfront condominium. The minute we closed on the condominium, we also bought a 13-foot Boston Whaler, so that Charlie, who was fourteen-years-old, would have something to do when he came to see us. I asked him how much he had in the bank and he said he had $600. The boat cost $6000. I think he was shocked when I asked him to give me the $600 so that he could be a 10 percent owner of the boat. Boy, did he take care of that boat.

This was the best thing I ever did because it made Charlie have skin in the game. After teaching him the ropes, he would bring his friends and they would go all over Long Island Sound and around Manhattan Island, which was a trip that I had made many times as a 10 to 15-year-old. The boat would actually go 40 mph. Manhattan Island is 12 miles long from the north end to the south end, so it is about a 25-mile trip around the island. Doing the trip from Stamford, CT, added about 20 miles. From Stamford to Hell Gate, it was a 20-mile trip down Long Island Sound, under the Throggs Neck Bridge and the Whitestone Bridge, past LaGuardia Airport through Hell Gate, and either north up the Harlem

River or south, down the East River. The route depended upon whether you were doing a clockwise or counterclockwise circumnavigation of the island. The views were spectacular, the currents were swift, and you really had to watch the weather. This was the beginning of Charlie's experience of following in my footsteps.

6
The Roaring 80s

Growing Thornton Tomasetti

THE KEY TO SUCCESS IN consulting engineering, especially structural engineering, is to be able to relate to architects, listen to architects, and collaborate. This was quite easy for me to do in light of the fact that I understood the ultimate aim of architects – to achieve a synergistic relationship between the aesthetics, the architectural systems, and structural systems. Many architects have commented that very few structural engineers can pick up a felt tip pen or a sharpie and instantly start drawing during an initial meeting. I could do this.

It took 20 years, from 1961 to 1981, for Thornton Tomasetti to get into the position of designing tall buildings in the over-40 story market. In 1978, the big boom in Atlantic City gambling casinos started. The first casinos were renovations of old hotels. Then came Bally's. With a net income of $1 million a day, $365 million a year, Bally's hired Cushman and Wakefield, one of the largest commercial real estate brokers and project consultants, to produce a 1.5 million square-foot, four-story gambling casino in 13 months. The architectural firm, Skidmore, Owings & Merrill (SOM), got the contract to design the building, even though SOM was not known for its gambling casino design experience. Most gambling casinos were designed by architects who basically had relationships with the gaming community. They generally produced ugly buildings that made a lot of money.

Bally's realized that in order to deliver the project within 13 months, they needed the experience of Cushman and Wakefield to pull together the whole team and deliver the project. Tom Imperatore and Tom Hays at Cushman and Wakefield led the effort. Turner was the contractor. Skidmore Owings & Merrill had other thoughts about who the structural engineer should be, but Tom Imperatore told them to hire Thornton Tomasetti. We got the job and I took the lead with Jay Prasad, Udom Hungespruke, and Dennis Poon. Seven days after we attended the first meeting, Cushman and Wakefield and Turner ordered 7,000 tons of steel. It was a wild ride, but we produced a 1.5 million square-foot casino with the potential to construct a 40-story building on top. This meant we were designing the foundations and columns for a yet-to-be designed tower to the casino.

Several years later, when the tower was built, the developer picked a hotel architect whose fees were the lowest in the industry. The only way this architect could accomplish this task was to hire a student to produce the design of the exterior façade. The architect laid out the hotel and really didn't care what it looked like. Because we had designed the foundations for the future tower and had the inside track about what went into the ground to support the future tower, the architect then contacted us to become the structural engineer for the job.

The only problem came when the architect told us that our fee would be one third to one half of what it should have been. We gave him our fee, which was three times what he planned to pay his usual subservient structural engineer. He had no choice but to hire us. We never did another project with this architect because he had no respect for structural engineering or any intention of allowing the structural engineer to make a profit. As the Bally's project proceeded, Tom Hays and I became very good friends.

Tom, his wife Sonia, and their three children, Thomas Jr., Brooks, and Lisa lived in Pleasantville/Armonk near my family. Our children attended school together. Tom took his trips to Atlantic City on a small commuter plane almost every week. One morning about 8 a.m., my phone rang and it was Lisa Hays telling me that her parents had just been killed in a plane crash attempting to take off at LaGuardia Airport. What a jolt this was. We did whatever we could to assist the family. Thornton Tomasetti later hired Tom Jr. as an intern. When things settled down, Tom Imperatore called me and said he needed to replace Tom Hays at Cushman and Wakefield. I suggested George Feddish, who had started with LZA in 1961, for the job and he stayed with Cushman and Wakefield for at least 15 years. During this time, George continued Cushman and Wakefield's tradition of hiring Thornton Tomasetti for jobs, including many of Thornton Tomasetti's key people including Fruma Narov, Jay Prasad, Dennis Poon, and Udom Hungespruke. George is still a great friend of Thornton Tomasetti and a great friend of mine.

Tony Peters, Vincent Peters, Joe Peters, and Tom Imperatore, who were owners of Cushman and Wakefield with several other family members, were Merchant Marine graduates. They were all on liberty ships during World War II, essentially running the engine rooms on these ships. The natural thing to do in 1945 and 1946 when they came off the ships was to go to work in the boiler rooms and central plants of New York City's major office buildings. These men ultimately worked their way out of the boiler room and into the management of these buildings. While they were very strong on mechanical, electrical, plumbing, and fire protection, they were not strong on architectural and structural engineering. This is how Thornton Tomasetti fit in so beautifully.

In New York City, a person who looks after his or her friends is called either an Angel, a Rabbi, or even a Godfather. That's exactly what these Italian American Merchant Mariners working for Cushman and Wakefield were for Thornton Tomasetti. We performed miracles for them and they remained loyal to us for close to 30 years. I would say that if there was one entity that enabled Thornton Tomasetti to climb into the big time, it was Cushman and Wakefield, the Peters family and most importantly, Tom Imperatore.

Our first major project together was in 1981 with One Tampa City Center in Tampa, FL. As General Telephone and Electronics (GT & E) headquarters, it would become the tallest building in Florida. The architect was Welton Becket, who didn't know or want Thornton Tomasetti as the structural engineers, but Tom Imperatore of Cushman and Wakefield prevailed and we were asked to design the 42-story building. As a result of this project, Hank Brennan of Welton Becket, who wasn't familiar with Thornton Tomasetti, gave us all of his work for the next 20 years. His firm later became Brennan, Beer, and Gorman.

Hank Brennan of Brennan, Beer, Gorman (BBG) is one of the best architects I have ever known and is part of one of the best architectural firms that we ever worked with. Hank Brennan, David Beer, and Peter Gorman worked for Welton Becket, a Los Angeles-based architectural firm. The New York office of Welton Becket was extremely successful because of these three people who ran it. Over the years, I have seen egocentric, self-centered owners of architectural firms refuse to give up a fair share of the profits to the key employees, so consequently, the key employees leave. This was the case with Hank, Peter, and David, who left Welton Becket and went on to form BBG.

Back in the 60s, the New York Association of Consulting Engineers did a study of the rotating preferences of architects about structural engineers, mechanical, electrical, plumbing and fire protection firms (MEP & FP). The thrust of this study was to see how many years it took for an architect to decide to change structural engineers. The average was about 10 years for structural engineers. The same question was asked about MEP engineers and the average was two to three years. Since architects have a much better understanding of structures than they do of MEP, the interaction between the structural engineer and architect is much more proactive, collaborative, and creative than with MEP engineers.

One Tampa City Center was the first of many jobs with BBG. After working together on several projects, I invited Hank Brennan to go sailing on Long Island Sound with Joe Denny and several other friends and clients. Joe and Hank hit it off immediately and BBG was selected for the Society Hill Sheraton Hotel in Philadelphia and Thornton Tomasetti became the structural engineer for the project. As a result of inviting Hank Brennan sailing, BBG started to do the BBG Regatta out of Manhasset Bay Yacht Club and I was invited to bring my sailboat to join the other boats owned by Peter Gorman and other clients. This was the start of a long relationship of sailing with BBG's clients. One of the things about business development is that repeat business is much easier to get than new business. Client sailing ended up being the vehicle for the growth of the company to be a geometric progression – network, network, network.

Thornton Tomasetti's next project was the American Motors Headquarters in Southfield, MI. It was a project with cost overruns. We worked with Tom Imperatore and Mike McCambridge of Cushman and Wakefield. This project led to the 42-story Continental Center at South Street Seaport in lower Manhattan – our first structural steel frame high-rise office building in New York City. Swanke, Hayden, and Connell were the architects on this job and they became our lifelong friends.

In the mid-1980s, as the 100th anniversary of the Statue of Liberty approached, Richard Seth Hayden of Swanke, Hayden, and Connell called to invite me to join an oversight committee to review the work of the French and the American team participating in the design work leading up to the restoration of the statue. My work consisted of several meetings in New York and several meetings in Washington, DC, with a group called the French-American Committee for the Restoration of the Statue of Liberty. The leader of the oversight committee was George White, the architect of the Capitol. One of the benefits of my involvement in this project was a trip to Paris to meet the French artisans and engineers working on the unique aspects of the skin and frame of the statue. It was the one and only time that I took the Concorde from Paris back to JFK Airport – what an experience! The photo on the back cover of the book by Dan Cornish captures the unique relative size of the nose of the Statue of Liberty as it relates to a six-foot, three-inch tall person. A close look at the picture reveals where the rivets connecting the skin to the frame on the Statue's face have failed.

In 1983, Cushman and Wakefield was selected to deliver United States Steel's second major project in downtown Pittsburgh, PA – the Mellon Bank building. Originally called the US Steel Headquarters building, this building was 55 stories high. Cushman and Wakefield selected Brennan, Beer, Gorman Architects and Thomas Imperatore told BBG that Thornton Tomasetti was a structural engineer. David Beer was the designer and Frank La Susa was the project manager for BBG. For Thornton Tomasetti, the project was handled by Richard Tomasetti, Len Joseph and Abe Gutman. The project utilized an exterior steel skin, which participated in the control of lateral drift and wind forces – a very innovative approach that US Steel wanted to adopt to show the versatility of steel. Mellon Bank later purchased the project and today it is known as Mellon Bank's headquarters in Pittsburgh.

As a result of our work in Pittsburgh, Thornton Tomasetti also won the Consolidated Natural Gas Tower in Pittsburgh with Lincoln Properties and Forest City Ratner's innovative Hilton Hotel at the convention center in Pittsburgh, using the Forest City Dillon precast concrete panelized structural system. We also did 5th Avenue Place in Pittsburgh, with Stubbins Architects out of Boston, and later helped design PNC Park, the Pittsburgh Pirates ballpark. In business development, one thing always leads to another. Staying in touch with everybody that's in the decision-making position of a company always leads to more work.

During this period, we also built an additional trading floor in an existing building at the Chicago Board of Trade in Chicago, IL, working with three architectural firms, Murphy/Jahn; Shaw & Associates; and Swanke, Hayden, Connell Associates Architects. Tom Imperatore insisted we join the team and we ended up with the project. Our work with Murphy/Jahn led to a later $300 million 25-story expansion of the Chicago Board of

Trade building. Once Helmuth Jahn learned of our technical capabilities and creative design collaboration, we were awarded the Northwest Atrium Center, the United Airlines Terminal #1 at O'Hare International Airport and many other projects. The United Airlines terminal is one of my top six favorite buildings, along with the American Airlines Super Bay Hangar, Petronas Towers, New York Hospital, and UBS Warburg Swiss Bank in Stamford, CT.

One of the lead managers on the Chicago Board of Trade project was Roy Harlow, an architect by training, who worked for Cushman and Wakefield. Roy had been the project manager for GTE at One Tampa City Center. He befriended Jay Prasad and Udom Hungspruke from Thornton Thomasetti, which led to a very important break for Thornton Tomasetti. I realized we were getting jobs we never would have been considered for 10 years earlier.

Sometime around 1984, Cushman and Wakefield (C & W) was acquired by Mitsubishi and Rockefeller Center Properties. Mike McCambridge of C & W was placed in charge of all renovations and construction at Rockefeller Center – a very large complex in Manhattan. The original structural engineering firm, which designed all of Rockefeller Center, somehow had the idea that since they did the original design, they owned all the drawings, specifications and calculations and therefore Rockefeller Center had to use them to do the structural engineering work. Mike consulted his attorneys, who said that Rockefeller Center actually owned the information. Mike called me and we negotiated a deal and from that point on Thornton Tomasetti did all of the structural engineering work in connection with renovation, alterations, and new construction at Rockefeller Center.

One of the lessons I learned over the years is that if you do a good job for your clients, they keep coming back. Relationships between architects, engineers, contractors, sub-contractors, and real estate companies go on for years. The entrenched companies that are involved in these relationships are very hard to knock out. Cushman and Wakefield discovered Thornton

Tomasetti and we knocked out prior relationships with structural engineers on many projects. Every project they brought us in on, we redesigned a poorly conceived structure, integrated the structure with the architecture, and brought the project in on time and within budget. Cushman and Wakefield began to get us involved in everything they touched. Regardless of the size of the project, whenever Cushman and Wakefield called, we jumped.

Thornton Tomasetti had very aggressive business development habits in the late 70s. We were chasing an athletic facility at Lehigh University, directly through the University and through some Bethlehem Steel people. Lehigh has always been supported by the steel industry. We were told that their architect was Ed Reidel's firm, Warner Burns Toan and Lunde (WBTL), an outstanding university planner and international hotel design firm. Although we didn't get the Lehigh job, we did several hotels and office buildings with WBTL in Jeddah and Medinah, Saudi Arabia, and in Yaoundé, Cameroon. Ed Reidel and I became instant friends. Known as "Uncle Ed" to my family, Ed was a gem of a personal friend, a great father, and a great confidante. Like most architects, Ed ran an architectural firm, but not in a business-like manner. We are still absolutely great friends, but our working relationship came to an end after the Jeddah Hilton.

One day while sailing with Ed out of Westbrook, CT on my sailboat, we were anchored out on Long Island Sound. The Jeddah Hotel had led to significant pressure on me from my partners, because our firm wasn't getting paid for the services we were providing. As we consumed a second Tanqueray martini, Ed and I got into a fairly loud argument. I said to Ed, "So who is the Captain?" He said, "You are!" I then said, "Shut up. We will never work with you again, but I want to maintain my friendship with you. Our friendship is more valuable than losing money with you on your international hotel projects." Ed and I have maintained that friendship to this day. I first played golf with Ed at the Producers Council Golf Outing at Sleepy Hollow Country Club in Westchester, an affiliate of the American Institute of Architects. The event promoted interaction between architects and product suppliers. Richard and I were not very good golfers. We played with Ed and got to know him a little bit, but it wasn't until I met Carolyn in 1979, that we started hanging out with Ed, who lived in Manhattan. Ed, his girlfriend Mary Ann, and Richard frequently sailed and dined out with us.

Ed was one of my clients whom I invited to go sailing. We used to tow people in light wind, on dock lines with a loop on the end over the stern of the boat. On one trip we were towing a lawyer and a fish bit his toe. We joked about professional courtesy – shark to shark! On the fourth Male Bonding sailing trip, my brother Bill and his son Michael came. They were on my boat along with Ed, Harry Armen and several others. Ed has a wonderful sense of humor. My brother is a little stiff. So for most of the trip, Ed was mischievously pretending to hit on my brother. Throughout the trip, Ed tells Bill that he is attracted to him and that he is very interested in him, but doesn't let on that he is kidding. Bill doesn't know what to do. On one of the last nights, Ed essentially proposes marriage to my brother Bill. Ed will do anything to become my brother-in-law. Quietly, Bill's son, Michael, is listening and observing. He finally makes the following statement, "Ed, I just don't think I could ever call you mom." This was one of the funniest memories from all of the Male Bonding cruises.

Ed also spent a lot of time with us in Connecticut with my mother Evelyn. The two became quite good friends. My mother was an amazing, warm, and friendly person. So one night, years later, after we moved to Maryland, Ed was visiting us and proposed marriage to my 90-year old mother. He'll do anything to be my stepfather. I think he was trying to get an inheritance.

World Financial Center at Battery Park

In the early 80s, Olympia and York, a Toronto-based developer, bought the Uris Empire in New York City, a company that owned at least 10 major high rises in Manhattan. Olympia and York announced that they were going to build their first buildings in New York and hired Roy Harlow from Cushman and Wakefield as a project manager. The new building project would include four office buildings and a lot of support space, retail space, parking and a Winter Garden – called the World Financial Center at Battery Park City. Uris owned a lot of old buildings and Olympia and York bought them all. Nat Stein was the person at Uris who had all the relationships with New York structural engineers. Unfortunately, we were not one of them.

One day Jay Prasad walked into my office and asked if I wanted to have lunch with Roy Harlow. Roy had just moved back to New York and he was interested in getting Thornton Tomasetti on the short list for the structural engineering of World Financial Center at Battery Park. Cesar Pelli would be the architect. Jay set up a lunch at Nanni, a great Italian restaurant on 46th Street between Third Avenue and Lexington Avenue. Tom Imperatore always ate there and it had tablecloths with felt tip pen drawings by Der Scutt, the architect for Trump Tower, hanging in the restaurant. Architects love drawing on tablecloths or backs of envelopes. So Jay and I had lunch with Roy. The lunch went well and he asked if we wanted to get on the short list to be considered for the structural engineering on an eight million square foot project. Even though Thornton Tomasetti had only just entered the high-rise steel office building business in 1981, with One Tampa City Center, of course the answer was yes. We were the ninth firm to be added to the short list. The list included the usual New York City suspects, plus Severud Associates and Gillum Associates.

The first step in the selection process to be considered for the structural engineer job was to be interviewed by two of Olympia and York's trusted Toronto-based structural engineers, Mordy Yolles and Rollie Bergman, who came to New York City alone to interview the nine firms. Richard Tomasetti, Abe Gutman, and I met with Mordy and Rollie for about two hours. Abe Gutman and Rollie Bergman hit it off immediately on the details and Richard and I hit it off immediately with Mordy Yolles on the big picture. The next step was that they would invite three firms to come to Toronto to meet Keith Robert, executive Vice President of Olympia and York's. Along with Thornton Tomasetti, Severud Associates and Gillum Associates were interviewed. Richard and I had to fly back from Toronto on the day of the interview to attend the ENR Award of Excellence Dinner that night. We learned the next day that we had won the job.

Because I was quite busy at the time, Richard Tomasetti and Abe Gutman, backed up by Udom Hungesbruke and Paul Lew, oversaw the World Financial Center at Battery Park project. Thornton Tomasetti did two of the four towers, the Winter Garden, the north and south bridges over West Street, and a gateway that links the buildings. Yolles did the other two buildings. We worked out common agreement on details, framing systems, and the approach to the project, which made for a very successful association of the two companies. Through our work on the World Financial Center at Battery Park and the Winter Garden, our reputation in New York City was significantly enhanced.

During the World Financial Center at Battery Park project, I developed a relationship with Norman Kurtz of Flack and Kurtz. It became apparent that Flack and Kurtz had contacts and a network that were very compatible with Thornton Tomasetti. Flack and Kurtz were mechanical, electrical, plumbing, and fire protection engineers and we were structural engineers. We were complementary to each other and could market each other's firms and clients. Norman Kurtz also developed a very strong relationship with the Cesar Pelli group.

Norman called me from his ski house in Colorado and suggested that I make an immediate call to Philip Isaacs and Hans Jensen in his Australian office. I made the call, befriended Philip and Hans and we ended up with the 50-story Chiefley Tower in Sydney, Australia, for developer Alan Bond with Kohn Pederson and Fox Architects based in New York. Alan Bond fielded the team that took the America's Cup from the New York Yacht Club, of which I am a member.

In a similar manner, later, Norman Kurtz had a meeting in Toronto with Jay Cross, the president of the Toronto Raptors. Jay Cross asked Norman who the best structural engineering firm was for sports facilities, including arenas, and he said Thornton Tomasetti. Jay Cross called me and arranged for a meeting in Toronto. Tom Scarangello and I went up and met with Jay Cross, who became one of Thornton Tomasetti's best clients. Jay Cross was like our long-term relationship with Cushman and Wakefield. The old adage came true – if you do a really great job for a client, they will keep coming back. Norman and I had a fantastic relationship.

My personal relationship prevailed with Norman Kurtz until his death at 69 years of age. He was a champion tennis player. In fact, at age 40, he won the gold medal in the Maccabiah Games in Israel. I would never play tennis against Norman – he was out of my class. Norman Kurtz was a gentleman to the end.

The Paper Bridge

In the early 80s, I received a call from Alex Cooper of Cooper Robertson, a prominent architect and urban planner in New York City. He said to expect a call from Ogilvy and Mather (O & M) advertising agency. O & M wanted to know who could design a paper bridge to carry a 12,000-pound Mercedes-Benz truck for a commercial they were making. Cooper Robertson said to call Charlie Thornton with Thornton Tomasetti and within a couple of days I received a call from an O & M account executive asking us if we could do the job. Of course, I responded, "Yes!"

The bridge was to be constructed and used in a commercial for an early Apollo mission. As usual the whole process was behind schedule. The client would be International Paper Company. We met with the advertising agency and requested the specific technical information we needed to design the bridge. The bridge was to be 32 feet long, 10 feet wide, and 4 feet deep, and it needed to be able to carry a 12,000-pound truck. The bridge had to be completely fabricated of paper and Elmer's glue with no staples and no steel reinforcing.

Paul Lew, Joe Thelen, and I came up with the concept, performed the calculations and arranged for the bridge to be constructed in Chicago. The deck was made of tubular cardboard elements that would normally be used to roll carpet. The actual structure of the bridge looked like an egg crate – a three-dimensional triangulated cardboard bridge. Since the schedule was tight, we acted quickly – designing the bridge, arranging for the paper to be shipped to Chicago, and building the bridge. As a precaution, we load tested the bridge in the plant where it was fabricated with a strategically located load pattern, simulating the wheel loading of the truck. To our surprise, the deflections were much larger than we expected.

We called International Paper Company's engineers and found out that it took about three weeks for the cardboard to cure – in other words, to get up to strength. So, we put the bridge on the truck and sent it to the Valley of Fire National Park near Las Vegas, NV. We arranged to rent a crane and placed the bridge over a 32-foot wide small canyon. O & M hired Roger Ward, a daredevil racecar driver to drive a truck. Fortunately, with the dry air and low humidity of the Las Vegas area, the bridge dried out and the paper came up to strength.

The day of reckoning arrived. Roger Ward walked over to Dr. Joseph Thelen and for obvious reasons, started questioning his credentials. After all, Roger was about to take a leap of faith and drive a 12,000-pound truck over a paper bridge. Apparently, Joe convinced Roger that he was for real and they proceeded with the test.

The final commercial showed a happy-go-lucky truck driver driving across the desert. As he approached the bridge, he stopped with a puzzled look on his face. He then moved onto the bridge and the tubular "corduroy-like" road deck actually crushes and the paper goes from being circles to ellipses. The truck continues across the bridge and into the sunset across the desert. While a distinguished voice touts the ability of paper to be strong, light, fireproof and durable, a Chinook helicopter flies in and picks up the bridge and flies away. The bridge ended up in a museum in North Carolina for a while. I suspect it eventually ended up in a dumpster.

This was another example of Thornton Tomasetti doing projects that were out of the box.

599 Lexington

In the early 80s, we became the structural engineer to design 599 Lexington, the firm's first major tall building in Midtown Manhattan. The building was 50 stories high on a site just south of the Citicorp building on Lexington Avenue and 49th Street. Our other Midtown tall building was America's Tower, designed by Swanke Hayden Connell. We got the job for 599 Lexington through Joel Weinstein's relationship with Edward Larrabee Barnes, who was selected by Cadillac Fairview, a Canadian developer. After commencing the project, the Canadian developers ran into some financial issues and Cadillac Fairview bailed on the project. Boston Properties, with Mort Zuckerman and Ed Linde, picked up the project. We had to be re-qualified to be the structural engineers, since Boston Properties had never worked with us. The good news was that I already knew the new person running the Boston Properties New York office, Robert Selsam. I had met Robert who was with New York City Planning Commission, in about 1975, after we won the contract to be the prime engineer/architect on the Roosevelt Island Tramway for the New York State Urban Development Corporation. Upon meeting Robert again for the Boston Properties project, he said to me, "Look, I know you guys can do the job, but you need to go to Boston and you need to meet with Ed Linde, the president of the company. Ed is an MIT engineer and you need to convince him that you're the right firm to design the building."

I flew to Boston, spent the day with Ed Linde and his team and they voted to continue with us on the project. This was the beginning of long-term relationship with Robert Selsam. The two very tall buildings at Times

Square and Broadway were developed by Boston Properties and Thornton Tomasetti provided structural engineering on both projects. Richard Tomasetti and Aine Brazil continued the relationship with Robert. In 2000, when I moved to Washington and acquired James Madison Cutts, I found out that James Madison Cutts actually did most of Boston Properties work in the DC area.

During the period we were working on the 599 Lexington Avenue project, I invited Robert Selsam to go sailing on one of our client sailing trips. We became very good friends and after working together on projects, I invited him on our Male Bonding sailing trips. He went on several trips and enjoyed them all. He continues to do trips with some of the remnants of the MB Sailing gang. Robert took to sailing like a duck takes to water, bought a Grady White and used it at his home on Fire Island.

As I later prepared to leave New York, I asked Robert if he would replace me as Chairman of the Salvadori Center – a middle school program that co-teaches with middle school teachers, using the built environment to help students overcome math phobias. He accepted and took the program to new heights.

Client Sailing

I always loved the excerpt from the poem, "Sea-Fever" by John Masefield:

"I must down to the seas again, to the lonely sea and sky,
And all I ask is a tall ship and a star to steer her by;
And the wheel's kick and the wind's song and the white sail's shaking,
And a grey mist on the sea's face, and a grey dawn breaking,"

A Lassie comic book I read around age eight inspired my fascination with sailing. I loved the idea that Lassie was on a 20-foot wooden sloop and was fascinated with the Marconi sloop's mainsail and headsail.

Although I spent my childhood on boats, they were mainly powerboats. I did have a sailing kayak and Grumman sailing canoe. Several years earlier, a young woman architect at the American Airlines Hangar project, Barbara Hearn, took my family sailing on a Morgan 25. She and her husband, John Gardner, named it Fata Morgana, after the character in King Arthur's Roundtable, Morgan Lefay. We sailed out of Darien, CT, over to the sandpit on the North Shore of Long Island. This was my first time on a sloop. It was a very foggy morning. We left the marina with zero visibility and John Gardner navigated his 25-foot sailboat through the fog until the sun burned the fog off. We arrived an hour or two later in the sandpit, anchored and had a meal, and sailed back on a clear afternoon. I was hooked instantly.

In early 1983, Carolyn and I bought a C & C 25-foot sloop in Essex CT, near Old Saybrook on the Connecticut River. We labored over what to name it. As we continued to discuss names, Carolyn said, "You know, in business, in marketing, and in branding and positioning a company, an elegant solution is just a clever way to describe a great algorithm, software program or a solution to a problem." So we named the boat "Elegant Solution." It came time to sail "Elegant Solution" to Stamford, CT, to our slip, which we had purchased with our three-bedroom condominium at Schooner Cove in 1981. At this stage of my sailing career, which started in 1979, I only had about two years of experience sailing.

I asked Robert Nachemann, a colleague at Thornton Tomasetti, to join me for what was supposed to be a one day sail. Carolyn drove us to Old Saybrook, on the west side of the Connecticut River. Bob and I got on "Elegant Solution." Although I was unfamiliar with the boat, I said, "So what." We motored down the Connecticut River taking advantage of the current and tides and entered Long Island Sound. We raised the sails and spent the whole day sailing toward Stamford. When we got near Bridgeport the wind died. We turned on the engine and were motoring when we emptied the first small tank of gasoline, which drove the small Atomic 4 gas engine. Most of the powerboats that I had had did not require you to vent the gas tank. This one did, so when we switched tanks, and the engine died after a half-hour, I thought we were out of fuel. Even though the tank was still essentially full, we were not actually getting fuel because later we learned that we had not vented the tank. We put up the sails and moved slowly toward Stamford in light air – with no wind.

I was monitoring the weather on the VHF radio. Gale force winds were predicted about 10 p.m. that night. It was obvious that we would not to make it to Stamford. I called the Coast Guard and explained the situation. The Coast Guard was located at Eaton's Neck, directly across Long Island Sound on the Long Island side. The Coast Guard responded by asking us whether our sailing gear was intact. I responded "yes" and they responded, quoting some Coast Guard regulation, that if we were on a sailing vessel with an engine and our engine dies, but our sailing gear was intact, we were not in a position to be assisted by the Coast Guard. This was before Towboat USA and other private towing services. They asked me to stay in touch and to keep them abreast of our progress, which I did.

I called Carolyn on the Marine operator and explained that we were going to go into Cedar Point Yacht Club on the Saugatuck River in Westport, CT. This was where I actually sailed the CSY 37 Viking on years of client sailing. As you would expect, the wind died. Fortunately, there was a very small current, but at high tide, which was coming within an hour, the current would switch and we would probably have been swept back out into Long Island Sound with no wind and no engine. With a flashlight shining over the stern, I was able to visually ascertain by the tiller and the rudder that we were moving north in the direction of Cedar Point Yacht Club.

By now it was dark, there was no wind and we slowly but surely worked our way to the entrance to Cedar Point Yacht Club. We had both the main and the headsail up and were probably moving at less than one knot. Of course, just as we were nearing a slip, the wind came up and we accelerated toward the dock with no engine. I released the main sheet. Bob was up front and I yelled to him to drop the headsail, also known as the Genoa, but he said he didn't know how, which came as no surprise. I ran up and put my feet out and took the shock as the boat rammed the dock. Carolyn was standing on the dock watching it all as we came in. We left and went back to our condominium in Stamford. Bob went back to New Jersey. The next day, I picked up the boat and sailed the rest of the way home alone.

People always ask me how to learn to sail and I always respond—just go sailing. As Nike says, there is only one way to learn – "Just do it." Just like in life, whenever you go out on a sailboat, you learn something new. If you don't try something new, you never ever get out of your box. I try never to repeat anything I have done in the past.

I learned a number of things while sailing. First, I learned why they call the boom, the boom. As I had learned, "Elegant Solution" had a rather unreliable, small Atomic 4 gas powered engine. One day, we were sailing out of Schooner Cove and a fairly strong thunderstorm arose on the western horizon. We headed into Stamford Harbor without an engine, so I had to leave the sails up. I should have dropped the mainsail and only had the jib up, but I didn't because that would have required me to head up into the wind. This would have slowed us down and I wanted to get up the East Branch of Stamford Harbor to a more protected spot than out in the middle of the harbor.

In order to drop the mainsail, you have to be headed into the wind. We were going downwind, meaning the wind was behind us. As I recall, it was Kathy, Charlie, Carolyn and me. Charlie knew that I had to make a 180-degree turn, but Kathy didn't know this and unfortunately she was standing up. As I swung the boat into a 180-degree turn, the boom came across, bumped into Kathy and knocked her over the side of the boat. Fortunately, she was able to hang onto the lifelines and we pulled her back in. Usually this kind of an event, getting hit by the boom, happens when you do an accidental jibe. As I always say, the only way to learn to sail is to just do it, make some mistakes and hope that everyone survives.

Early in our relationship Carolyn, Diana, Kathy, Charlie and I were sailing in Narragansett Bay. Captains, in the early stages of sailing and lacking their own confidence, tend to yell at their crew. Once you know what you are doing and you gain confidence and experience, you stop yelling. On this day, we were sailing and it seemed to my crew like I was yelling at them. Actually, because the wind was blowing so strong, I had to raise my voice in order for my crew to hear me. So Kathy and Carolyn were up on the deck doing something with the mainsail and I yelled at them. Kathy in no uncertain terms said, "Dad, if you don't stop acting like an asshole, we may never sail with you again." I never yelled again.

My good friend and client Richard Hayden of Swanke, Hayden, and Connell had a boat on Tortola in the British Virgin Islands. I researched Caribbean sailing and learned that the easiest place to learn to sail in the islands was the Sir Francis Drake Passage out of Tortola. So, I charted his Mariner 38 and Carolyn and I went sailing for a week. It was terrific. I was extremely cautious and made sure that every night we anchored in the right place and set the hook to avoid dragging at night. Everything was going well until the last night. We anchored in Trellis Bay, adjacent to the airport, and went into Tony Snell's Last Resort on the small island in the harbor. We arrived late, which was a no-no because the bay was crowded with other boats. I found a spot and didn't set the hook.

We went in for drinks and dinner and got back to the boat around midnight. Sometime around three in the morning, we were hit by a 60-knot squall. I knew it was coming, so I was sitting in the cockpit with the engine running, facing the wind. All of a sudden all of the boats around me seemed to be going forward. We were going backwards. I revved the engine to 3000 RPMs and it was still dragging. Finally, as usually happens in a squall wind, the wind abated and we were able to move forward, pick up the anchor, and proceed up to the other end of the harbor and drop anchor. This time, I set the hook.

Sailing in the British Virgin Islands is very easy, but I wanted to sail in blue water, so in 1991, I got the bright idea that we would do a father and son sailing trip. I selected St. Lucia – Marigot Bay. I decided that we would sail from St. Lucia up to Martinique and back. It was a hard sell to convince my spouse that I should go sailing in January and leave her in the wintry north, but calling it a father-son sail seemed to work.

I chartered a boat and Charlie and I, my brother Bill and his son Michael, Ed Reidel and his son Jonathan, and Bob Liss and his best friend's son Gary Lesch went on the trip. This was the beginning of the January all-male sailing trip each year. We did 13 Male Bonding sailing trips and even had a website called MB Sailing. I also did books for each trip, including the crew, itinerary, and photos and memories. Each crewmember filled out a profile which included such things as their favorite beer, favorite music, snoring index, and how they got along with architects. One of my favorite comments on one of the logs was by Greg Chiu about his yachting experience, where he noted, "I saw 'Mutiny on the Bounty' 20 times, and I've read 'Billy Budd.' I was also a Boy Scout where I learned to tie many different types of knots that would be really useful when we tie up the architect." There is always a fun banter between engineers and architects!

I started with all my best friends like Bob Liss, Harry Armen, Ed Reidel, Joe Denny, and my brother Bill. I then extended it to the clients and friends whom I enjoyed spending time with, including Robert Selsam, John Chirco, Brian Donnell, Henry Michel, George Pavarini, Wayne Hubbard, Ed Dean, Chris Rojahn, Ted Von Rosenvinge, Michael De Chiara, Alan Pifko, Steve Baldridge, Craig Smith, David Prevatt, and Dean Stephen. The trip started with two or three boats and grew to as many as seven boats, with probably a seven-man crew, so we were up to about 50 people. On the early trips, Ed was the cook, but as time went by, Harry Armen pushed him out of the kitchen. Harry would wake up early and get to the kitchen before Ed. Phil Wilde, Henry Michel's son-in-law, was famous for his cinnamon rolls, which were a part of every trip he was on.

Henry Michel grew Parsons Brinkerhoff from a 500-person New York City-based consulting engineering firm to a 5,000-person international engineering powerhouse. He was my hero and a great mentor. We met at SAME and Civil Engineering Research Foundation (CERF) functions in NY and Washington, DC. During 1979 thru 1995, Henry mentored

Richard and me as to how to grow the firm and brand. Henry, in his late 60's at the time, came on three MB Sailing trips with his son-in- law. On one of the MB Sailing trips, Henry was taking his term at the helm off the west end of St. Marten in sustained winds of 35 knots in 10 to 12-foot seas. Steering a 50-foot boat under full sail under those conditions takes a certain amount of energy and effort, so Henry requested permission to vacate the helm position and Wayne Hubbard took over and hit a button. Henry asked, "What's that?" Wayne answered that it was the auto-pilot. He said, "Nice guys!" Henry treated Kings, Sultans, Sheiks, and taxi drivers with the same respect. He was a diplomat and an engineer. He passed away at 72 years of age.

MB II and MBIII were one boat 10-day trips. MB II was St. Martin to Nevis to St. Kitts to Antigua to the southern end of Guadalupe – a long trip.

MB III was another 10-day trip to Grenada up to Tobago Keys and back. Don Ercole, Jeff Boak, my son Charlie III, and Ed Reidel were on this trip as well. MB IV was the first multiple boat trip – every trip after that was multiple boats. One trip was sailing around Antigua for seven days, and others included trips in Belize and the Bahamas. Another trip was a British Virgin Islands, Tortola through the Spanish Virgin Islands, past Culebra and on to the East End of Puerto Rico. This was a Sun Sail Charter and the French-designed boats were horrible, unlike the Moorings Charter Bvi boats which I chartered many times. Our MB XIII sailing trip in St. Maarten was the last trip we did. These trips were amazing networking opportunities and were an important part of Thornton Tomasetti's success in business. We got to know our clients on a personal level, as we did with our earlier hiking trips, which helped to foster longstanding relationships that benefited all of us.

7
My Second Family

New York City Living

WE STAYED AT SCHOONER COVE until 1984, at which time Carolyn and I moved into a condominium on 22nd Street between Third Avenue and Second Avenue. The girls finished school in Armonk and were off to college. Diana and Kathy, the two oldest, were great athletes and good students and were self-motivated. Charlie was less self-motivated and was living with his two grandmas and a grandpa, who was not very communicative. With the two girls gone, he was as focused as he should have been. Years later, he said it was like the movie, "Charlie and the Chocolate Factory," where the boy, Charlie, lives in a small house with both sets of grandparents. He remembered the grandmothers having tea when he got home from school each day.

Carolyn encouraged me to get Charlie into a private school and he ended up going to Cushing Academy and did very well playing three sports a year. It forced him to become a pretty good skier. In fact, he was number two on the downhill ski team. His roommate, who was number one on the team, was from Switzerland. After one year at Cushing Academy, his sisters said he got a personality. He graduated from Cushing and was admitted to the University of Hartford.

Meanwhile, Diana went to Gettysburg College to play varsity volleyball without a scholarship and Kathy went to Babson and graduated number one in her class in accounting. As a senior, Kathy organized all the interviews for the big six accounting firms on her campus. I told her, "That is very smart – you will know all of the recruiters." She said, "Duh?" My kids always were one step ahead of me. During this time, Carolyn, who had been a professional woman all of her life and knew about appropriate business attire, took Kathy to Orchard Street in New York City to buy her first suits for her first job. Carolyn's insights with the children were a huge help to me.

Becky Arrives

In the spring of 1986, we were having dinner with Bob Peckar and his wife Maxine. By this time, Carolyn and I had decided we wanted to have a child of our own. On this particular evening, Carolyn was telling Maxine about how rough it was to conceive a child. Bob overheard the conversation and told us to call Stanley Michelman, a lawyer who specialized in adoption, and adopt a baby. He gave me Stanley Michelman's phone number. I called him and arranged for a consultation for $300 in Rockland County, NY. Carolyn and I drove up to Rockland County and spent about an hour with him to learn more about independent adoption, which was legal in New York, but not in Connecticut, where you had to go through an adoption agency. During his career, Mr. Michelman was instrumental in arranging adoptions for more than 5,000 adoptive parents. His book, *Private Adoption Handbook*, provided guidance for hopeful adoptive parents.

He told us to take out advertisements in "Pennysaver" magazines with an unlisted number that would ring to our condominium on 22nd Street in Manhattan. The advertisement said, "White couple wants white baby – boy or girl. Couple treasures education and will provide a great environment for your child." We blitzed upstate New York with $1,500 worth of advertisements and on or about April 1, the phone started ringing.

The calls were interesting. The first phone call was from a grandmother whose daughter had to give up one child for adoption because of mental issues. She was pregnant again and asked if we would be interested. The answer was no.

Second call came from a woman in Buffalo, New York who had two children and was pregnant with her third child and her husband had just walked out on her. She was Catholic and did not believe in abortion. She asked if we would take her child. Even though we answered yes, she never called back.

Sometime in the end of April, a young woman named Rebecca called. She explained to Carolyn that she was over six months pregnant and had sat down with her mother a few days before to discuss the situation. The mother decided she really didn't want to take care of her daughter's baby. So Rebecca told Carolyn that she had seen several advertisements from our "Pennysaver" initiative and had made several calls. She asked Carolyn if she had time to answer some questions. Carolyn obviously said yes.

The first question was about whether we both valued education. Carolyn explained that she had a master's degree from UCLA and that I had a master's degree and a PhD from NYU in engineering. She then asked what we did for entertainment. Carolyn answered that we had subscriptions to the New York Philharmonic, the American Ballet Theatre, the New York City Ballet Theatre, the Circle in the Square Theatre, and the Roundabout Theatre. The last question was about what we did on weekends in the winter. Carolyn told her we skied in the winter and sailed in the summer. Carolyn thanked her for calling and asked her for her phone number. We tried to get her lawyer's number, but she was reluctant to give it to us. Carolyn was convinced that we were going to get a child. By mid-May, we had received no call. And then Stanley Michelman called one day and told us it looked like we were going to get a child. Rebecca's lawyer had gotten busy and forgot to call Michelman. Rebecca called and told us that she wanted us to be the parents.

We only had about three weeks' notice that we were about to adopt the child. I was the veteran of having three children, Carolyn was the novice, and she was beside herself. She wanted to buy furniture. I took the top drawer of my dresser, opened it, emptied its contents, and said, "Babies are small. We'll just put the baby in here and then we'll buy furniture." After we got the news about the adoption, I called my mother and told her that she was going to be a grandmother again in a week. She said, "Doesn't it take longer than that?" I told her we were adopting a baby. We hadn't told anybody we were seeking to adopt because we thought it would take a year. It took about four weeks.

Rebecca Carolyn Thornton was born on June 19, 1986 at Crouse Memorial Hospital in Syracuse, New York. After we received the call that Becky had been born, we chartered a two-engine, two-pilot plane at Westchester Airport and flew to Syracuse with a female attorney. Prior to our arrival at the Syracuse airport, we had called for a taxi to take us from the airport to the hospital and to take us immediately back to the airport. As we were driving into town in the cab, the three of us in the back seat were discussing the fact that we were picking up the baby. Upon arriving at the hospital, the lawyer told the cab driver to pull around behind the building and wait there until she came out. This was to avoid meeting the birth mother or whoever came out with her. The minute the lawyer got out of the car, the driver turns around and says to us, "First grandchild?" I couldn't figure the logic of that question! So he parked around the corner and the driver said he had to go in and use the men's room.

When the driver returned to us he was all excited and said, "I've seen the baby." Before I could stop him, he put the car in gear and drove right up to the front door of the hospital. So we are now parked right outside as the lawyer comes out with the baby. We have no option but to get out and say hello to the birth mother and her brother. The driver didn't really understand what was going on. We returned to the airport, got on the plane, and flew home.

The risk in adopting is that the birth mother had within a year to change her mind. We made it through the first month, which is the critical period when a mother can change her mind and reverse the adoption. Then we made it to the year period and Becky was ours to keep.

Art

At PS 69, my teacher, Mrs. Lash, noticed that I had a lot of drawings of comic book covers – mostly G.I. Joe type stuff. She asked me if I was willing to have my first one-boy art show on the fourth floor bulletin board. Also in grammar school, I had a notebook where I drew the map of every country on the planet. Each page was dedicated to a country. I showed the Capital city and other major cities and the rivers. To this day, I still have a great sense of geography and a great sense of direction, which really helps you find your way. My father had been an excellent draftsmen who was good at mechanical drawing. I taught myself how to draw.

In 1967, after I moved to Pleasantville, NY, I played around with acrylics, but not seriously and continued some painting through 1970. My real love of art was stimulated by the Clark Art Museum in Williamstown, MA. This museum was probably the best French and European Impressionist art museum in the United States. The collection of Monets and Toulouse-Lautrecs there is fabulous. There were also a few Winslow Homers and Frederic Remingtons. I also became enamored of the Wyeth family – N. C. Wyeth, Andrew Wyeth, and Jamie Wyeth.

In 1986, when Becky was born, we were living in Manhattan on 22nd Street between Third Avenue and Second Avenue. We spent weekends at our waterfront condo in Schooner Cove, Stamford, CT, and on Elegant Solution II, our Tartan 37-foot sloop. Initially, Becky took about a one-hour nap, so I started doing 45-minute paintings on canvas. There is really no reason why an abstract Impressionist or an Impressionist work of art should take more than 45 minutes. That is probably my attention span anyway. I started painting in the upstairs loft and really got into it.

Once Becky got over three-years-old, we went sailing every Saturday and Sunday – just the two of us. By this time, we had moved to our four-bedroom house in Dolphin Cove with the boat slip right in the backyard. It was very easy for me to depart and return, single handling the boat. We were located at the end of the lagoon, so I could make a 180-degree turn and bring the boat right in alongside the dock, essentially parallel parking it between the other two boats on our dock. I never needed any help from anybody. All my Tartan sailboats were rigged for single handling. What this means is that all of the running rigging—the halyards and jib sheets – all lead back to the cockpit. This enabled me to sail the boat without ever leaving the cockpit. This is the safe way to sail – especially while alone.

I would set Becky up in the vee berth in the bow cabin, with all her stuffed animals. As I motored out from the dock and raised the mainsail, I would look into the vee birth and she would already be asleep. If I sailed for one to three hours, Becky took a one to three hour nap. She always woke up when I turned on the engine. Obviously, Carolyn loved this as it gave her a break to do whatever she wanted to do. I got to go sailing and Becky took a nap.

When we adopted Becky, I remember telling Carolyn that since we had traded the C & C 25-foot sailboat for our Tartan 37-foot sloop and that we were scheduled to do a pursuit race from Stamford, CT to Northport Harbor on July 4, I fully intended to take Becky sailing on her first overnight sailing trip when she was just three weeks old. It was a down-wind start and Becky started screaming as we crossed the start line. Needless to say, we did not finish (DNF) and we were probably DFL – you can probably figure out what this means.

For several years, our Christmas card was actually one of my paintings. I painted pictures of both of my brother's boats. My favorite subject to paint were lighthouses. I liked the geometric form and the ability to give depth and shape by shading from white to gray to gray to white. I did several different paintings of the Stamford Light House – many of which hang in my children's homes. I also donated some Giclee prints of these paintings to silent auctions.

I continued painting through the 80s and into the 90s and really accelerated my painting when I moved to the Eastern Shore of Maryland. There are about 50 of my paintings hanging in our house. The talent for painting must run through the family, as my cousin Hugh "Mickey" Thornton is a talented artist in Lewes, DE.

Sailing and Engineering

One memorable story about sailing and engineering occurred in 1986 when Karl Sabbagh, a film producer, was in the process of doing a PBS special on Zeckendorf's Worldwide Plaza Building. This was at the same time that Karl was writing the book entitled, *Skyscraper – The Making of a Building.* I invited Karl, his assistant Nicola, and his camera crew to go sailing on Elegant Solution II. The purpose of the sailing trip was to show how sailboats under wind load act very similar to tall buildings under wind load. I always like to try and demystify engineering to people. The lesson is that the weight of the building and the width of the building, parallel to the wind, is how to figure out the resisting moment – the point a building can be toppled. If the overturning moment is smaller than the resisting moment, then the building can withstand the wind. The overturning moment or the tendency of the sailboat to heel or tip is very similar to that which is happening on a tall building. The difference is that as the boat heels and the center of gravity of the keel moves out toward the wind, keel weight times the distance to the center of gravity of the keel, is the righting moment. When the righting moment is greater than the overturning moment, you don't turn 'turtle' in the sailboat. Another way to say this is the mast does not end up horizontal. I loved using sailing to teach.

Yachting Parties

When we moved back to Connecticut in 1989 and purchased the house at 19 Dolphin Cove Quay, I noticed a large 80-foot long Trumpy Yacht at the west end of the cut at the exit of the lagoon where I kept Elegant Solution III. I walked over one day, when there was significant activity on board, and inquired as to whether the boat was available for charter. It was. So for about four years in a row Thornton Tomasetti charted the Trumpy Yacht "Enticer" and invited 60 to 70 clients and staff for an evening of cruising on Long Island Sound.

The yacht was actually large enough to have a sit down dinner. The menu was usually filet mignon or Châteaubriand served with vegetables and dessert. Prior to dinner, we served cocktails and fancy hors d'oeuvres to the guests. All but one of the cruises we had perfect weather, which allowed the guests to be up on the top deck, as well as inside the cabin. One of the other trips, we went down to Greenwich Harbor and cruised around Greenwich. On the other cruises, we would motor right across the sound to the North Shore of Long Island or to Rowayton and the Five Mile River.

One of the most memorable of the cruises was when a hurricane was brewing about 50 miles out in the Atlantic, off the South Shore of Long Island. Fortunately, on that trip, in light of the storm, we stayed behind the breakwater at Stamford Harbor and only experienced some rain. Since everyone was inside the cabin for cocktails and hors d'oeuvres and dinner, the party hit a critical mass and was probably one of the most exciting and fun parties we ever had. In the close quarters, everybody was mingling, conversing and having a great time.

The boat was finally sold and I had the daunting task to inform each and every guest who had been on the four trips that the boat was no longer available. People had really come to enjoy this annual event and we didn't want people to think they had been removed from the guest list. After this, we reverted to normal client sailing on my 41-foot boat. These sailing trips were one-day trips, generally on Thursdays. We left from Stamford Yacht

Club at three o'clock and sailed until six o'clock with 10 to 12 guests on board. We docked the boat and proceeded to the Stamford Yacht Club deck and watched the sun go down while having drinks and dinner. As with the Male Bonding sailing trips, these client sailing excursions provided a venue where I could entertain friends and clients. Playing golf with clients did nothing to impress them about my overall abilities, but while sailing, I could share with them my zest for living and risk taking, while also sharing my own self-confidence – important qualities when hiring a structural engineer.

More Sailing

In 1996, we moved up to a new Tartan 4100, and named her Elegant Solution III. She was a beautiful, flag blue-hulled Tartan 41 foot sloop which displaced 22,000 pounds with a huge cockpit and a very fat, wide, beam. She was a very great boat in heavy seas. We broke her in on the return trip from our four-week summer cruise to Maine in early August 1996. Carolyn and Becky had decided to fly home from Maine. John Chirco, Dick Harlan, and Lois and Jim Brown, and I were on the boat. Coming out of New London, CT, we headed west, we saw black thunderheads beginning to form. We dropped Lois and Jim Brown off in New Haven. Jim was a research professor at Yale and an IBM fellow at the Thomas Watson Research Labs and Lois was a friend of Carolyn's from Exxon office system days in Stamford, CT.

The three of us remained on the boat and came out of New Haven Harbor and motored down past the Connecticut shoreline in an absolutely flat sea. As we approached the islands off of Norwalk and the Saugatuck River of Westport, CT, the storm hit. We saw it coming, we had plenty of warning. The genoa and main were secured with extra sail ties. We had jack lines from bow to stem on each side (port and starboard on the deck), we all had harnesses with tethers to the pulpit. We were all hunkered down within our dodger with all plastic glass in place around us with the Bimini above our heads. We felt protected.

The storm hit at about 2 p.m. I had never experienced a storm like this. The ozone wave of superheated air, distinctly perceptible by smell, hit us and within one minute, the 60 knot wind was on us. As we entered into the storm, Elegant Solution III performed beautifully. With the first and strongest gust, it required 3000 rpm on the 42 horse power Yanmar diesel engine to barely keep the bow into the wind. The bow went under with each succeeding wave. Water poured over the deck. We had six to eight inches of water pouring over the deck. About 3 minutes into the event, I looked down to the steps going into the cabin and there was water pouring over the floorboards. This was not a good thing for a sailboat in a storm. I turned the helm over to John Chirco and told him I was going down below to check what was going on. The first thing you do on a sailboat is to taste the water you see. If it is salt water, you are in trouble. If it is fresh water, it means that your tanks have just let go. When I tasted it, it was salt water. We were in trouble.

I came back up and took over the helm and continued into the storm, but the good news was the 60 knot winds had diminished to 50, 40, 30, and I was finally able to go back down and ascertain that the three bilge pumps were clogged with what looked like sand, or so I thought. I continued to explore what was going on while now standing in ankle-deep water on my beautiful teak and holly sole. I couldn't figure out how sand could be in the bilge pumps. I guessed this storm just roiled up the bottom since we were right off the Norwalk Islands in 25 feet of water. In retrospect, I should have gone much farther out into deeper water where the waves would have been less severe. But, this was another learning experience. As the storm passed, within 15 minutes, I went down and discovered that all of the filters in all of the bilge pumps were actually filled with sawdust, not sand. Evidently, as Tartan built all of the beautiful interior teak woodwork, they never blew out or vacuumed the sawdust from under the various and sundry panels and beds. This left the bilge pumps clogged with sawdust.

Next, I had to figure out how salt water got into the boat in such a large quantity. First, I realized that the hawse pipe at the bow, which allowed the anchor chain to enter and leave the anchor chain locker, had allowed water to fill up the locker and enter the bilge through a hole in the bulkhead. Second, I realized that I had inadvertently left open the through hull fitting in the vanity sink in the V-berth, which was in the upfront sleeping area. As we plunged under the water and the bow went ten feet below sea level, the hydrostatic pressure just blew water all over the V-berth – all over the mattresses, blankets, and bags. It took about three weeks to dry the place out.

We finally pulled into the dock at our home at Dolphin Cove in Stamford, CT, at 5 p.m. that day. We had succeeded in experiencing, maybe, the perfect storm. But, more importantly, the Tartan 4100, Elegant Solution III, had gotten us home. She was a true elegant solution!

PART TWO – ELEGANT SOLUTIONS

8
Sports Facilities

"Those who dream by night in the dusty recesses of their minds wake in the day to find that it was vanity: but the dreamers of the day are dangerous men, for they may act their dream with open eyes, to make it possible."
—T. E. LAWRENCE, *SIX PILLARS OF WISDOM*

A RELATIVELY YOUNG ENGINEER NAMED David Geiger started a firm called Geiger Berger Associates and cornered the market on tensile fabric, inflated and non-inflated roof systems, used in about half the domed stadiums in the US. I always felt that the liabilities of tensile fabric roof would ultimately come back and bite the designer. Dave Geiger's projects in Minneapolis, MN, Pontiac, MI, and St. Petersburg and Gainesville, FL, all ended up with partial collapses during heavy rain or snowstorms. The lawsuits took their toll on David Geiger, and he passed away in 1989 at a relatively young age.

I learned over the years that the way to capture a market segment is to stay on top of where the action is, where the projects are, and who the major players are. With the death of David Geiger, there was an opportunity for Thornton Tomasetti to enter the market designing sports facilities. First, we had to follow who was selecting structural engineers to design long span arena baseball ballparks and football stadiums. The major players were Hellmuth Obata Kassabaum (HOK) Sport, Ellerbe Becket and HNTB – all located in Kansas City. With Thornton Tomasetti's long-span roof experience, we were perfectly suited to introduce our integrated and collaborative constructability approach to these long-span sports structures.

In 1989, Sally Handley, Thornton Tomasetti's marketing person, and I went to a Society of American Military Engineers (SAME) annual meeting in Kansas City, MO. Prior to the trip, I made arrangements to visit Chris Carver and Rick deFlon with HOK Sport and Eric Piper with HNTB. I knew that they were each desperately in need of a new structural engineering firm and a new way to construct these long span spectator type facilities. As a result of the meeting, Rick deFlon and Chris Carver asked Thornton Tomasetti for a proposal on the Chicago Bulls and Blackhawks Chicago Stadium, now called the United Center. We won the job and the first meeting with Rick was intentionally set in Chicago at Comiskey Park, which was under construction.

There were some structural issues with the structural engineering on Comiskey Park. HOK Sport was in a quandary because they couldn't get the City of Chicago's structural plan examiner to give them a building permit. I agreed to help, but insisted that we would not take over someone else's structural design. When I returned to the office and met with Tom Scarangello and Fruma Narov, I found out the original engineer had resigned from the project. Tom and Fruma, upon hearing that HOK Sport wanted Thornton Tomasetti to take over at Comiskey Park, wanted to see my Superman shirt. This was the beginning of Thornton Tomasetti's work in sports facilities across the country, including the Anaheim Mighty Ducks Arena in Anaheim, CA; the Pepsi Center Arena in Denver, CO; and converting the Anaheim football and baseball stadium to only a baseball stadium for the California Angels.

In 1996, Jay Cross, president of the Toronto Raptors basketball team, called me at the suggestion of Norman Kurtz of Flack & Kurtz, a mechanical, electrical, and plumbing firm on the World Financial Center at

Battery Park project. Norman suggested Thornton Tomasetti as the structural engineer for the new arena in Toronto. Tom Scarangello and I flew to Toronto and met with Jay Cross. Jay was very direct and stated that although we couldn't be the engineer of record for a project in Canada, he suggested we interview three Toronto-based structural engineers in New York City to show them we were really in control of the selection process. We interviewed three firms, including Yolles. There was no doubt in my mind that Yolles was going to get the project with us. This was our first sports project with Jay Cross. Later, Thornton Tomasetti did the Miami Heat arena, when Jay was president of the Miami Heat; the Jets stadium in the West Penn Central Yards, when he was president of the Jets; and lastly, the combined Jets/Giants MetLife football stadium in New Jersey.

I became very friendly with the Yolles. In fact, as the Raptors project progressed, I said to Andy Bergman, Rollie Bergman's son, that I wanted to have lunch with Mordy Yolles and Rollie Bergman. During that lunch, I asked them why they recommended Thornton Tomasetti to Olympia & York for the World Financial Center at Battery Park. Their answer was very interesting and telling, and provided a good lesson for Thornton Tomasetti and all engineers. Mordy and Rollie told me that Thornton Tomasetti was the only New York firm that showed the two of them any respect. So the moral of the story is to understand that everybody in the process is important; not only the people who you think are important. Every one of those other firms made the wrong call. They told me that one engineer even had a putting green in his office and he was putting golf balls into the hole during the conversation.

Since I had not worked on the World Financial Center at Battery Park, I really didn't know anyone at Cesar Pelli's office. Tom Morton from Cesar Pelli was running the project and Gregg Jones and Jon Pickard from Cesar Pelli were also very involved in the project. They loved what we were doing and really liked our team, including our creative engineer, Paul Lew. Paul is a genius. Although he was then and still is a little eccentric, sometimes wearing two different colored socks with two different shoes, Paul had six degrees in engineering and architecture. I met Paul Lew in 1967 when he was standing in the lobby of Lev Zetlin Associates with a six-foot long double layer cable bridge model he had built at Tulane University as an engineering student. When I walked into the office that morning, I said, "I am Charlie Thornton, who are you and what can I do for you?" He said, "I am Paul Lew and you need me!" So, I invited him in and explained that we were not hiring at that time. He went on to join the firm, Iffland Kavanaugh and Waterbury, but we stayed in touch. When things picked up, we hired him. The key to running a successful engineering firm is to fit round pegs in round holes. Some engineers can do business development, some can make cold calls, others are great on the production of contract documents, and some, like Paul Lew, are great with concepts. We used engineers in the roles where they excelled.

Near the end of the World Financial Center at Battery Park project, Paul Lew came to me one day and told me that Cesar Pelli's architecture firm had a lot of projects coming up and that nobody in Thornton Tomasetti was really doing business development in New Haven with Pelli's people. He gave me Tom Morton's name and said I should go up and see him because he was "one of my kind." Tom was a naval officer before he went to Yale Architectural School and was more of an administrative manager than a designer. I went up to New Haven to see Tom Morton at Cesar Pelli, where I also met Gregg Jones, Jon Pickard, Fred Clark and Cesar Pelli. This began my relationship with Cesar Pelli that led to us collaborating on the Petronas Towers and many other projects. The moral of this story is to never stop selling in your business. Just because you did a great job with someone, doesn't mean they will call you again.

9
Hospitals and Institutional Buildings

IN THE EARLY 1990S, THORNTON Tomasetti began working on some challenging and complex engineering projects that would strategically position the firm in the world market. These included expanding New York Hospital between 68th and 70th Streets in Manhattan, expanding the Lincoln Laboratory at Massachusetts Institute of Technology (MIT), and building UBS Warburg's office towers and trading facility. These projects were important because they were large, complex and high profile.

The New York Hospital expansion project, which was completed in 1994, consisted of an 800,000 square-foot addition of a 15-story air-rights structure spanning the six-lane FDR Drive between 68th and 70th Streets in New York City, as well as renovations to the existing 23-story hospital. FDR Drive, also known as the East River Drive, is a highway that runs from the tip of Manhattan all the way up to the Triborough Bridge and basically is the edge of Manhattan along the East River. Because of traffic constraints and maintenance of traffic requirements from the DOT's, you can only build over the drive between the wee hours like midnight to five in the morning. You need to be out of the way when the rush hour starts.

Fritz Reuter, the project manager, is right up there in my top 10 client list. Fritz and his team assembled the best design and construction team that I have experienced. Unlike many other owners, Fritz and the New York Hospital wanted to do it right!

The architectural team on this $450 million project was HOK and Taylor Clark Architects. Lehrer McGovern was the construction management company, led by Cary Colton on the overall management and Ken Hiller on the structural side. Thornton Tomasetti was the structural engineer. Syska and Hennessy was the MEP Engineer. Canron was the steel fabricator and erector, led by Larry Davis.

The project started with conventional thinking, fearing New York City union jurisdictional issues. As time went by, Aine Brazil, now Vice Chair of Thornton Tomasetti, and I developed a relationship with Hank Winkelmann, the lead architect on the design side for HOK. Hank is the most brilliant hospital architect I have worked with. His ability to listen to the consultants and collaborate and integrate everybody's ideas into a solution is magical. Hank is from St. Louis and he commuted weekly into New York City to work on the project. I believe he is still doing this to this day – that is how good he is.

The project started going and we decided that there had to be a better way to build it. At every meeting I sat next to Hank and Aine. As the project evolved, Aine and I convinced Hank Winkelmann that the only way to build this building was to prefabricate the deck in New Jersey in 800-ton modules that could be placed on a barge and lifted off the barge every night with the thousand-ton crane. Next, we talked to Ken Hiller and Canron and Larry Davis and the deal was done. The end result was a masterpiece of collaboration, synthesis and integration. Fritz, as the owner of the New York Hospital expansion project, is a shining light in this industry.

The weather cooperated and the 7 sections were transported up the East River on barges to the site. Cranes on floating barges lifted 950-ton sections into position – breaking all records for weight lifted by cranes. Some of the challenges in the project included seismic considerations, site constraints, and code requirements. Despite these challenges, the project was completed on time and roughly $4 million under budget.

The other project we were working on during this time was the expansion of the Lincoln Laboratory at MIT in Lexington, MA, a leader in surface and solid state physics and materials research. The laboratory's innovation and application of advanced technology for surveillance, identification, and communication play an important role in our nation's defense. Because of this, the floor structures of this five-story laboratory building had to be designed to minimize floor vibrations and accommodate highly sensitive research, such as the semiconducting laser and infrared laser radar for high precision pointing and tracking of satellites.

The next project, one of my favorite buildings, was the UBS Warburg in Stamford, CT, which consolidated all of the Swiss Bank Corporation's operations in one location. The design and production architect was Skidmore Owings and Merrill (SOM) and the contractor was Turner Construction Company. The project was constructed in three phases – a 15-story office tower, a nine-story trading facility, and several levels of parking. The most significant function of the complex was its trading operation, a 44-foot by 240-foot area, which made it the world's largest clear span trading floor.

In the book, *Expressing Structure: The Technology of Large Scale Buildings*, writer Virginia Fairweather describes the ceiling in the trading space of the Swiss Bank building as including "mechanical, acoustical and lighting systems within a dramatic series of waves, following the building's subtle nautical themes." SOM's Mustafa Abadan credits Thornton Tomasetti for creating suspension of the monumental staircase that is, "aesthetically pleasing, almost invisible cable connection, inspired by sailboat design."

This job required a high degree of accuracy and minimal construction tolerances, as well dealings with the newly revised seismic requirements for the expansion joint between the office tower and the parking/trading facility.

10
More Tall Buildings

Miglin-Beitler Tower

IN THE MID-1980S, DURING A visit to the office of Harry Weese, a famous Chicago architect who designed the Washington DC Metro, I saw some experimental designs that his firm was working on for a tall building in Chicago. I offered Thornton Tomasetti's structural engineering help, and Paul Lew and I came up with a 2500-foot tall building for the site, called the Chicago World Trade Center. Weese promoted the project in major newspapers as if it were real. "Engineering News Record" did a big article on it in 1983 entitled, "How Tall?" The article put us on the map.

During this time, although Thornton Tomasetti was doing very special projects like Comiskey Park, the Bulls and Blackhawks facility, United Center, the United Airlines Terminal at O'Hare International Airport, the Northwest Atrium Center at the Northwestern Railroad Station, and the Chicago Board of Trade expansion, we really didn't have presence in high-rise concrete residential buildings and composite core and steel floor office buildings. We had developed a relationship, however, on many of these projects with a structural engineer, Eli Cohen of Cohen Baretto and Marchertas (CBM) and with Lee Miglin and Paul Beitler of Miglin Beitler, real estate developers, who had developed some of these buildings. In fact, the first time I met Eli Cohen, we were sharing engineering on some projects. Eli completed multiple 70-story concrete residential buildings along Lakeshore Drive and along many other downtown streets within the loop in Chicago.

In 1989, CBM was completing the structural engineering for Miglin Beitler's Madison Plaza III, a 60-story building at Madison and Wells in Chicago, with Cesar Pelli and Gregg Jones. Paul Beitler and Lee Miglin wanted to develop the world's tallest building - The Chicago Star Spire. The perception of many industry people in Chicago was that CBM could easily design a building up to 70 stories, which would be about 700 feet, but a 2,000-foot tall building was perceived to be beyond their capabilities. So CBM and Thornton Tomasetti were teamed to produce the structural engineering on this project called the Miglin Beitler Tower.

The building was to be 2,000 feet tall with a very small footprint – the typical floor was only 120 feet by 120 feet. Because of the Chicago Zoning Ordinance, we would be able to put some outrigger walls that would increase the structural footprint to 140 feet by 140 feet. Bob DeScenza with Thornton Tomasetti started the project before he moved to Chicago to run the Chicago office. Len Joseph and Udom Hungespruke took over Bob's role with me. We had designed a spectacularly efficient building, which made all the headlines because it was going to be the tallest building in the United States and in the world.

Chicago has the best concrete in the US because of the Dolomitic Limestone aggregate with two suppliers capable of producing up to 18,000 PSI concrete. We worked with Dick Halprin of Schal for the construction, Cesar Pelli and Gregg Jones on the architecture, and HKS on the production, and ESD for mechanical, electrical and plumbing, as well as my heroes, Clyde Baker of STS for geotechnical and Peter Irwin of RWDI for wind engineering studies. Years later in an "Engineering News Record" article written about me, I commented on Clyde's precision in taking the mystery out of deep foundations, stating, "I go to sleep at night because of

Clyde Baker." Years later, Clyde would help me solve one of the most difficult pier foundation problems I would ever have while constructing the twin Petronas Towers in Malaysia.

We convinced the local concrete suppliers and STS consultants to perform some high-strength concrete drilled piers, also known as caissons, in the US. I was on cloud nine – we were going to be structural engineers on a 2,000-foot tall building that would dwarf everything else. Although this project was completely designed, the real estate recession of 1990 killed the project and it was never built. We continued, however, to work with Cesar Pelli on projects such as the Mayo Clinic and Norwest Tower in Minnesota, an airport terminal competition in Kansai, Japan, and a multipurpose sports facility competition in Saitama, Japan.

Petronas Towers

I developed my relationship with Jon Pickard at Cesar Pelli when we had worked together on the airport terminal competition in Kansai airport in Japan and almost won it. At the same time I was working with Gregg Jones on a proposal for a multipurpose sports facility in Saitama, Japan. We did not win that one either. We did, however, win the Petronas Towers project in Kuala Lumpur, Malaysia. To this day, Paul Beitler still tells me that he paid for us to learn how to do that type of building with the Miglin Beitler Tower and we successfully used all of the ideas on Petronas Towers. Paul was right and he paid us a lot of money to learn how to do it.

Since the Petronas Towers were only 88 stories and 1,500 feet tall, much shorter than the Miglin Beitler Tower, of course three years later I could say "absolutely" to the Prime Minister when asked about whether we could build it or not. In Cesar Pelli's book about the Petronas Towers, he attributes part of the success of the project to the "brilliant scheme developed by Charlie Thornton."

When we were selected to design the Petronas towers in Kuala Lumpur, Malaysia, we did not realize that they would become the world's tallest twin towers. The first trip to Kuala Lumpur was in November of 1991 with the entire team. Upon our return, we commenced our first phase of design generally called schematic design. We decided to look at five distinctly different structural solutions for many reasons. The primary reason was that Petronas was the Malaysian National Oil Company and their principal customers were Korea, Taiwan, and Japan. All of these countries are more familiar with using steel for tall buildings than with using concrete. Based upon our experience on several tall concrete projects, we determined that concrete would probably be more competitive pricewise. There was also a 45 percent import duty on fabricated steel, which was fabricated in another country.

We developed five schemes and submitted our report in October 1992, approximately 11 months after our first visit. Our report, dated October 22, 1992, addressed the question on the part of the President of Petronas asking why we were proposing a concrete system. Traditionally and historically, the tall buildings of the world, including Sears Tower, the Twin Towers of the World Trade Center, and the Empire State Building have all been structural steel framed projects. All of these projects had very large base dimensions relative to the height and therefore were efficiently solved using structural steel. In the case of the Sears Tower and the World Trade Center twin towers, a tubular solution with very closely spaced columns around the perimeter was used in order to efficiently utilize structural steel framing. These two buildings were also completed in the early 70s before the advent of high-strength concrete.

Recent experience in the United States has shown that as buildings get more and more slender with height to width ratios over 7, concrete structures become more efficient in resolving the lateral loads on the project, as well as allowing the designer to more easily minimize accelerations, drift, and motion perceptibility to the occupants within the building. As a result, recent buildings over 70 stories have been utilizing concrete and mixes, composite concrete, and steel systems. The shared experience in 1983 of Cesar Pelli and Thornton Tomasetti with the 1100-foot tall Norwest tower in Minneapolis, MN, confirmed our opinion that concrete is superior to steel, both structurally and cost wise.

Thornton Tomasetti completed two sets of 100 percent contract documents; one for the steel system and one for a composite concrete system. The exterior tube solution spaced approximately 3 to 3.5 meters on center around the perimeter on the 1.3 million square-foot tall project. The concrete solution was $6 million less than the steel solution. This was in the market when steel and concrete were comparable and generally competitive. If it were a 40-story building where wind drift, acceleration, and motion perception were not as critical, steel would have been equally competitive with the concrete.

We ultimately selected a concrete core in concert with a circular exterior soft tube, utilizing 16 concrete columns on an about 150-foot diameter plan form. The plan shaped selected by the architect fit beautifully with a 16-column perimeter tube. The views were spectacular and the column spacing was much greater than the prior steel structures. The schemes, which all used composite steel framing, deck and concrete, were as follows:

Scheme A - Structural steel core structural steel cylindrical tube system

Scheme B - Concrete core with structural steel cylindrical tube system

Scheme C - Structural steel core with concrete encased composite steel cylindrical tube system

Scheme D - Concrete core with concrete encased composite steel cylindrical tube system

Scheme E-- Concrete with core concrete cylindrical tube system

Working collaboratively with Cesar Pelli, Flack and Kurtz, Lehrer McGovern Bovis, and Thornton Tomasetti, we developed the preferred system – Scheme E. We priced all five schemes and jointly educated our Malaysian client as to why we were doing the approach we were proposing. I said to my team, which consisted of Bob Descenza, Len Joseph and Udom Hungespruke, backed up by a large team of other really key players from the company, that my job was to keep the client happy and to keep the job and their job was to produce an exceptional engineering design for an exceptional building.

I actually told them that we needed to educate our client, because I predicted the Prime Ministers of Korea, Japan, and Taiwan would call the Prime Minister of Malaysia and say "Are you crazy – you are allowing these crazy guys in New York to do a concrete scheme for the world's tallest building?" Our client has to have all the cards in his hands to basically protect the viability of our composite system. Extenuating circumstances on this project included a technology transfer. The entire team brought high-strength concrete, simple fabricated structural steel, fabricated locally, and a very sophisticated stainless steel façade, fabricated locally, to Malaysia. There was no way that an all-steel scheme could possibly compete with our composite scheme. I know for a fact that those phone calls took place between heads of state and we prevailed. We came up with an elegant solution, far better than anything that anybody else could come up with and we won. The project was topped out in 1996 and completed in 1998 - twin 88-story buildings accomplishing a floor every two days. One tower was built by a Japanese-controlled consortium and the other was led by a Korean-controlled consortium. It was a marvel to watch the competitive spirit between these two teams racing for the sky.

The 88-story Petronas Twin Towers in Kuala Lumpur City Centre represented a tremendous achievement in structural engineering design. They achieved the status of the tallest buildings in the world, their stainless steel pinnacles reaching 1,588 feet in height. The Skybridge at floors 41 and 42 connected the towers and created a visual gateway to a new major public park nearby. The complex also included a 5,143 car underground garage, a seven-story retail facility and an acoustically isolated 864-seat concert hall spanning over the project entryway.

In the 90s while I was focusing on the Petronas Towers in Malaysia, Richard, with Dennis Poon, concentrated on the first tall building in China. Richard and Dennis also focused on Indonesia and worked on many very large mixed-use projects in Jakarta, Indonesia. Although I started the ball rolling on Taipei 101 in 1989, by the time it really got going, I had moved to the Washington, DC area and Richard and Dennis Poon took over and did a spectacular job. We call it teamwork.

It wasn't by design that Richard Tomasetti and I tended to focus on different aspects of the firm and different projects. We gravitated to the things that we liked to do or allowed the other person to focus on what he liked to do. It was a great marriage or partnership or whatever you want to call it. On the national scene, I was sort of 'Mr. Outside,' meaning my projects were a wide variety, ranging all the way out to the West Coast. Richard tended to focus more on the East and in particular on really establishing Thornton Tomasetti's reputation in New York.

So in 1995 when we started the 10-year transition, I charted my path to end up in Washington for the last five years. Since Richard is two years younger, as of 2005, he had 7 more years in his position as major stockholder and chairman. Richard focused on the New York Building Congress, the New York real estate community, New York AIA and some charter schools with Richard Kahan. Richard rose to be chairman of New York Building Congress, chairman of the New York Building Congress Foundation and he started the Thornton Tomasetti Foundation. When Richard was nominated by the New York AIA to become an honorary member of AIA, it was mostly because of his fundraising efforts to allow New York AIA to achieve their goals of having a home on LaGuardia Place, right adjacent to the downtown NYU campus.

The Royals

On one trip, I arrived at Kuala Lumpur Airport and was told that my seat in first class was not available. It didn't really matter, but I asked why and was told that members of the Royal Family were on board, so instead of sitting in seat 2B on my way home, I sat in seat 2D. I got on the plane sitting in the second row on the right-hand side and on the left-hand side in the first row was a Scotland Yard inspector with the bulge under his jacket and right in front of me were Sarah Ferguson, Duchess of York, and her daughters, Beatrice and Eugenie, who were both about three or four-years-old.

By the time we reached cruising altitude, Eugenie and Beatrice were both sitting on my lap and I was showing them pictures of my daughter, Becky, who was about the same age. Fergie reminded me of every Irish strawberry blonde girl whom I knew when I went to school in the Bronx. I think she was a great mom. She stayed awake with the kids all night while their governess was snoring.

Because there was speculation about Fergie going to Bali with a boyfriend from Texas, the paparazzi had just previously photographed her at the airport in Indonesia in her black-and-white print dress, with the two little kids hiding behind her, under her dress. When I got on the plane, she had on the same dress. When we landed in London I had a time warp. All the tabloids had pictures of her on the cover in the same dress, playing the mother hen role. It took me a couple of seconds to figure out that the speed of light is faster than a 747 and that we'd been on for a 15-hour flight. So the wire services basically wired these photos to all of the tabloids – with Fergie splashed on the cover of them all.

I changed planes at Heathrow and on my next flight I told the flight attendant who was in the same seats on the last flight. She said Fergie was a commoner and I asked her what that made both of us. She told me how they felt about the royal family. It was a fun flight.

Taipei 101

This is a good example of how long it takes to cement relations, which ultimately lead to a successful contract. Asia is all about trust and relationships. In 1989, I was invited by the Taiwan Society of Architects and

Engineers to participate in a lecture series in several cities in Taiwan, including Taipei, Kaohsiung, and Taichung. During this one-week trip, I met Dr. Shaw Shieh, one of the organizers of the series and a professor and a consulting engineer in Taiwan. He was the chairman of Evergreen Consulting Engineers in Taipei. While traveling with him and the rest of the entourage, we became close friends. About that same time, Gyo Obata of HOK, one the largest US architectural firms, asked me to get involved in a project in Kaohsiung. It was at this point that I met Chris Cedargreen, an architect at HOK in St. Louis. HOK was not happy with the structural engineer on the Tuntex project in Kaohsiung, so they asked Thornton Tomasetti, including me and Len Joseph, to get involved in redesigning this large 70-story project.

It was at this point that I learned that there are very tight relationships in Asia between famous US engineers of Taiwanese background and the banks, insurance companies, and the steel and cement industries. We developed a really clever scheme that HOK loved. Len and I were ready to fly to Hawaii to meet with this engineer and HOK, when we got that call from HOK, which said that the engineer had arranged with the chariman of Tuntex for HOK to be fired if I came to Hawaii for this meeting. So that was the end of that one.

But in the process, I became aware of C.Y. Lee and C.P. Wang. C. Y. Lee was a Princeton graduate and C.P. Wang had been at Washington University in St. Louis, MO, and had spent some time with HOK. C.Y. Lee and C.P. Wang also worked extensively with Dr. Shieh. Within a short period of time, I met with all of these people in Taipei.

Starting in 1991, the trips to Kuala Lumpur occurred about four times a year for me. I couldn't stand Narita airport in Tokyo, so I would do anything to avoid a layover there. It became quite easy to depart from Kuala Lumpur and stop in Taipei for a day or two with C.Y. Lee and C.P. Wang and then continue back to the USA. In this process I became very friendly with C.Y. Lee, C.P. Wang, Eric Lin, Harris Lin, and their mother, a very wealthy widow who wanted to develop a very tall building in Taipei.

Shortly thereafter, the Lins, C.Y. Lee, and Dr. Shieh arranged for me to present a world's tallest scheme to Dr. Lee, the President of Taiwan, who held a PhD from Cornell. Although it took 15 years, these meetings lead to Taipei 101, our firm's second world's tallest building in 2006. Patience is a virtue.

11
The ACE Mentor Program

BACK IN 1961, WHILE IN my senior year at Manhattan College, I won a Society of American Military Engineers (SAME) scholarship equal to $100. SAME traditionally gave scholarships to seniors in college and I was one of the recipients that year. The scholarship amount was the weekly starting salary for an engineer in 1961. I actually bought an engagement ring with that money. Today, that same scholarship is about $1000 per student and well over 100 scholarships are given out annually.

After I received my PhD, I started to get involved in professional activities in New York City. Besides the American Society of Civil Engineers and the New York Association of Consulting Engineers, I became very involved in SAME, because the organization was one of the few that admitted it was established for networking and business development. Its purpose was to enable military officers, such as division engineers and district engineers in the Corps of Engineers, to learn who the movers and shakers are in each community. Since engineers in these positions generally stay less than three years, SAME was a means for these frequently moving Corps of Engineers officers to get to know local architects, engineers, and contractors. I started attending monthly meetings and quickly was assimilated into the organization.

In order to work your way up the chain of command in the New York Post of SAME, you had to get involved in the system running the annual black-tie scholarship dinner on the first Saturday night of November at the Waldorf Astoria. The ascendancy chain consisted of becoming a vice president, then vice-chairman of the SAME Dinner Dance, then chairman of the Dinner Dance, then president of the New York Post of SAME, and ultimately a director of SAME National. For me, it also led to me becoming a Fellow of SAME and a Golden Eagle, the highest honor that SAME bestows.

As a major contributor to SAME and the New York Post officer, I always attended the special Friday night dinner at the Metropolitan Club with Carolyn. This event was attended by about 100 people, all seated at one long table with spouses seated across the table from each other – officers and gentlemen and their lovely ladies. It is very traditional in the military for the host to toast the lovely ladies.

Carolyn and I had become friends with General Paul Kavanaugh and his wife Marjorie. They were sailors and had a sailboat. Carolyn and Marjorie were sitting together and Marjorie had kind of given Carolyn a little elbow and suggested that it was time for the women to toast the handsome men. Out of nowhere, after the gentleman toasted to the lovely ladies, Carolyn stood up in front of this rather staid conservative group and proposed a toast to the handsome men. Carolyn was younger than most of the women and she pulled it off very well. No one had ever done this before.

The SAME scholarship fund was run by Colonel Joseph Markle. Colonel Markle was a lawyer in the Judge Advocate General Corps during World War II and was friends with Max Urbahn, a famous New York architect who was also in the Corps of Engineers during World War II. Colonel Markle was insistent that the SAME scholarship fund award scholarships to seniors in college. He discovered me, as a scholarship recipient from Manhattan College, and singled me out as the poster boy who had "come full circle" and was now donating money for scholarships. This propelled me right to the top of the New York SAME Post.

In the late 80s, I prevailed on Colonel Markle that we should do four four-year scholarships for engineering students. I tried to convince the committee that giving the money to seniors was wrong. With a classic "if it ain't broke don't fix it" response, Colonel Markle rejected everything I wanted to do. At the end of the first four-year scholarship program, the SAME scholarship committee decided to discontinue the program.

So, fast forward to 1991, when Dr. Joseph Lestingi, Dean of Engineering at Manhattan College, called me and the other Board of Advisors to Manhattan College's Engineering School to tell us that the Board of Trustees was thinking of closing the Engineering School. The Board of Advisors acted quickly to reach out to minorities and women to interest them in architecture, construction, and engineering. We started with the New York City public high schools, creating summer internships in companies within the city. I developed the group further to include Steve Greenfield (Parsons Brinckerhoff), Ray Monti (Port Authority of New York and New Jersey), John Magliano (Syska and Hennessy), Ed Rytter (Chase Manhattan Bank), Robert Borton (Lehrer McGovern and Bovis), and Lou Switzer (the Switzer Group). In 1991, working with the Explorer Program of the Boy Scouts of America, the program began with one team of 30 students, mentored by volunteers from several firms. This developed into a mentoring program that would operate between October and May.

It was during this period, in 1992, that Richard Tomasetti and I made a $1.2 million commitment to endow Manhattan College's first faculty chair in engineering. As graduates of Manhattan College, we both felt great appreciation for our alma mater and how it prepared us for our success in engineering. We decided to name it the "Thornton-Tomasetti Chair in Civil Engineering," in honor of my father, Charles H. Thornton, Sr. (In Memoriam), and Richard's father, Angelo Tomasetti, Sr., both of whom had distinguished careers in the construction industry.

By 1993, Thornton Tomasetti had taken the lead and several companies had signed on to mentor students. Year-by-year, participation grew and by 1995, our group and Ed Rytter, head of construction and real estate for Chase Manhattan Bank and an old friend from American Airlines, realized that we needed to take our program to the next level. I organized a meeting at Ed's Metro Tech in Brooklyn. The meeting was attended by eight engineers, including Ray Monti, Chief Engineer of the Port Authority of New York and New Jersey, and Andy Paretti, Assistant Chief Engineer, Port Authority of New York and New Jersey; Lou Switzer, an African American architect with a very successful 100-person firm in Manhattan; Bob Borton, Lehrer McGovern Bovis; Robert Rubin, Postner and Rubin; David Peraza, Thornton Tomasetti; and John Magliano, Syska and Hennessy. At this meeting, we decided that we would continue our affiliation with the NYC School Construction Authority, but we would move into our own identity. We needed a name for our organization. We were all sitting around and Lou Switzer said, musing, "I think we should call it ACE because it was cool and an "ace" is somebody that is very good at what they do, such as a fighter pilot or baseball pitcher or cool kid from the streets of New York." We were all sitting there saying, "No, no it has to start with 'E' for engineering." But, after sitting there for an hour trying to come up with an acronym that began with an "E" we said, "You know, Lou, we need to call it 'ACE.'" In 1995, the independent nonprofit ACE Mentor Program, Inc. was formed – named by an architect.

From its onset, the ACE Mentor Program had a twofold mission: 1) to enlighten and motivate high school students towards careers in architecture, construction, engineering and related fields and 2) to provide the appropriate mentoring and scholarship opportunities for students so that they could become the future designers and constructors of our country. Finally, I would get the scholarship program I had hoped that SAME would implement years before.

The structure of ACE is that students would be recruited from both public and private high schools, with special efforts made to reach those, especially women and minorities, who might otherwise not be aware of the challenges and rewards of careers in the design and construction industries. Students selected for the program are divided into teams of 20 to 30 people. They work under the guidance of mentors from firms representing owners, architects, construction managers, and engineers (civil, structural, mechanical, electrical and environmental). Teams are also affiliated with a college or university. Teams meet at least 15 times during the course of the school year. Initial meetings include visits to the offices of the involved firms, where the scope of their activities is discussed and a tour of their facilities conducted. Each team then selects a design project that may require site acquisition, as well as the drawing of plans, the building of models, and other related activities. Students go through the entire design process, with the tasks they perform for their "clients" modeled on the real-life activities of their mentoring firms. Among the skills they learn are drawing to scale and estimating the cost of a job, skills that their mentors utilize in performing their daily professional duties.

In addition to the activities involved with these team meetings, the ACE Mentor Program sponsors field trips to colleges and construction sites. There is also a "How to Go to College" night where involved colleges explain their admissions' procedures and answer student questions; all ACE Mentor Program students and their parents are invited. At the end of the school year, there is a major culminating event at which all teams present their projects, much as actual design teams would present to their clients.

A Scholarship Luncheon, Scholarship Breakfast, and golf outings are held at each site to raise money for graduating seniors to help with the costs of attending college. Each graduating senior is given the opportunity to apply for an ACE Mentor Program scholarship and a committee comprised of Board of Directors, with input from team mentors, selects the scholarship winners. As of 2013, ACE has awarded more than $14 million in scholarships. Many of the mentors maintain contact with students after they have completed the program and gone on to college. Thus, mentors remain valuable resources for the students and often act as the important, caring adults needed to keep students on track when times become difficult. Many of the mentors themselves have indicated that they have grown professionally and personally from participation in the program.

After we incorporated the ACE Mentor Program in 1995, the program took off with the help of one of the earliest financial supporters of ACE – Raymond L. Smart of the Smart Family Foundation. This foundation supported charter schools in Connecticut with great success. I met Ray Smart through the Salvadori Center. We expanded the program to Newark, NJ and Stamford, CT.

One of the most popular and outstanding mentors for the ACE Mentor Program was John Chirco with Parsons Brinkerhoff. I met him while I was active in the New York Post of the Society of American Military Engineers (SAME). I was co-chairman of the committee to organize the SAME Annual Convention in New York City – the only time the convention was ever held there. Mary Ann Owsley, my administrative assistant at Thornton Tomasetti, assisted me with running the event. Since Parsons Brinkerhoff was very involved in the planning, they volunteered John Chirco, their transportation engineer to help. John and I became very good friends and he went on several of my Male Bonding sailing trips. He also hit it off immediately with Mary Ann. For some strange reason, John did not allow Mary Ann to tell me that they had developed a relationship. When I left New York in 2000, Mary Ann took a job with Microsoft in Bellevue, WA and John followed her and they bought a condominium together there. They are still friends today.

In about 1999, I was standing next to Nadine Post, an engineering writer with "Engineering News Record" *(ENR)*, at a final presentation night for ACE and I said that she should write a story in *ENR* about ACE. She said, "Charlie, I'll write a story when you take ACE to a national level." That's when I decided to take ACE national. Later, in 2000, when I moved to the Washington, DC area, we moved ACE to DC and expanded into many more cities. Additional people who were instrumental in helping

ACE to grow nationally were Richard Andersen, New York Building Congress; Jeff Levy, RailWorks Corporation; Rich Allen, Stantec Inc.; Charlie Bacon, Limbach Facilities Services; Tony Guzzi, Emcor; Preston Haskell, The Haskell Company; Linda Figg, Figg Engineering Group; Rick Kunnath, Charles Pankow Builders; Tom Gunkel, Mortenson Construction; Patrick MacLeamy, HOK Group; and Joan Calambokidis, President of International Masonry Institute. In addition to writing the article, Nadine nominated me for the *ENR* Award of Excellence in 2001 and I won the vote of the editors. Today, 40,000 inner-city high school students in 106 cities around the nation have been introduced to the challenges and rewards found in careers in architecture, construction, and engineering.

12
Thornton Tomasetti Acquisitions

Chicago

IN 1991, IT BECAME APPARENT to me that if Thornton Tomasetti remained a company exclusively based in New York City, we would never get our money out when it came time to transition the ownership of the company because it would be limited in size. To avoid that, we had to grow. If we stayed based only in New York City, the size of our firm could only grow to about 200 people. We had to get into other markets such as Chicago, Dallas, Los Angeles, and other population centers.

By 1992, we had laptops, cell phones, the Internet and all of the magical communication devices to manage our offices from a distance, so I suggested that we expand. Sometime around 1990, Eli Cohen of CBM, a Chicago-based structural engineering firm of about 13 to 15 people, called me and said that he was being approached by somebody else to buy his company and that he didn't like the firm. He told me that since we had been his associate engineer on many of his projects in Chicago and that he liked us, he would like Thornton Tomasetti to buy his firm.

I asked Eli to deliberate on a price and to get back to me. Six months went by and Eli called me back to ask me what was happening about buying the company. I told him he was supposed to call me and tell me how much he wanted. He still didn't have a price. In 1992, Eli called again and said, "I don't want to be bought by this other firm!" I finally decided that Larry Hine, Thornton Tomasetti's Chief Financial Officer, and I would go to Chicago to meet him for a day and talk turkey. I don't think we even told Richard what we were doing – he probably would have said no to our trip. We met with Eli, who was the sole proprietor of the firm, and laid out a game plan to purchase the company.

We asked Eli how much he expected to make, as well as his monthly payroll and expenses, including rent. He presented us a contracts analysis of all the contracts that he was presently working on, showing the percent completion of each contract. We told Eli we would guarantee him a fixed income, a five-year contract – the first three years as a full-time employee and the next two years as a part-time employee. We told him that we would take over his payroll and rent as of June 1, 1993. He could keep all of the accounts receivables that he had billed and could collect them on 'our dime.' We would get everything starting from the closing date that was billed on all his contracts. All it took to implement this approach was cash, which we had. We had to carry the Chicago payroll and expenses for at least three months because 120 days is average in collecting accounts receivables in the architectural/engineering business. Larry and I put together the plan listing all the benefits of the acquisition. We never listed any of the cons. Richard always thought of those.

The first thing I had to do before approaching Richard with the acquisition was to find somebody to relocate to the Chicago office and run it. We were never going to buy an office without having one of our trusted people in charge. So at a party at my home in Dolphin Cove, CT, I approached Bob DeScenza's wife, Margie, and asked whether she would like to move to Chicago. I'm not sure I had even talked to Bob about it, but I had determined that the most important person to make that decision was not Bob – it was Margie. So Bob and his wife agreed to go to Chicago. He had met Margie at Northwestern, so they had experience in Chicago. I now had everything in place except Richard's approval.

I'm not sure this is the first time I utilized this approach with Richard, but this is how it worked. Richard is an extremely intelligent, focused, non-multi-tasker. If you talked to him in his office, when he was thinking about something else, he didn't hear you. I realized that going in with a frontal assault, with the idea Larry and I had about acquiring a Chicago office, would not work. Instead, Larry Hine thoroughly prepared a financial analysis and I prepared a strategic plan as to how the acquisition could work. We packaged the proposal in a sealed envelope and handed it to Richard with the following instructions on the outside of the envelope, "Do not open until you arrive at 36,000 feet and you have had several Johnnie Walker Blacks on the rocks. We can talk later."

Upon his return, I would acknowledge that we knew he was back, but I wouldn't talk to him for two to three days because I knew he was preoccupied. Then, Larry and I would walk in and say, "What do you think?" Richard would typically say, "Let me give you all the cons. I'm surprised you guys didn't list the cons." Larry and I would look at each other and smile and under our breath would say, "Then you would have nothing to do." Subsequently, we would listen to the cons – we knew what they were – then we would counter with why they were not a problem. Richard would ultimately say, "If you both fail, then it's your fault, not mine." We used this approach over and over again with Richard since it was the only way to get him to focus on our ideas. Once Richard agreed – he was totally involved.

After we acquired Chicago, I spent a hell of a lot of time there. The only way that you can acquire other companies is to integrate the acquisition. You need to get to know everybody in the office and to dispel all their fears – especially the one that the "Wicked Witch of New York" has taken over the company. That's what we did. And nobody left, except those who we did not want to stay. Bob, Larry, and I, with a lot of help from others, grew the Chicago office from 13 people 125 people in three years.

The die was cast. We had our modus operandi and we knew how to implement it. From an accounting point of view, Larry Hine enabled these acquisitions to take place without acquiring any goodwill – which has big positive tax implications. We knew that these offices that we acquired were sleeping giants – the sole proprietor was pushing 65 and had gotten tired. He was firing the wrong people, not buying the right technology, and the company was basically becoming stagnant.

Eli's company, CBM, owned the developer market for structural engineering of high-rise reinforced concrete residential buildings in Chicago. They performed most of the plain-vanilla work. Thornton Tomasetti had already secured Comiskey Park, the Bulls and Blackhawks Arena, Northwestern Atrium Center, and the United Airlines Terminal at O'Hare International Airport. We had all the plum work and Eli had all the plain vanilla work, but large scale building projects. This was an acquisition made in heaven. It was a natural in my mind – the perfect combination of getting the plum work and not having to travel as much, since CBM was actually our associate engineer on a lot of these projects. We knew who we were dealing with and Eli liked us. We didn't know much about his staff, but they stayed as long as we needed them to stay and we weeded out the deadwood in a very professional and gentle way without angering anyone.

On the day we closed, Eli got a call from an architect named Jack Hartray that there was an interview at 3 p.m. for a project for a new radio station building at the Navy Pier – a very famous landmark on the Chicago waterfront. Eli asked us if we wanted to come to the meeting with him. So Jack Hartray announced to the owner of the radio station that today is a momentous day in the history of structural engineering in Chicago. He told them that Thornton Tomasetti and CBM had joined forces. If you want to span the English Channel, you want to hire Thornton Tomasetti. If you want to span the Chicago River, you hire CBM.

The next acquisitions that followed used the same game plan for Bridgeport, CT; Dallas, TX; and Tustin, CA. Part of my strategy of expanding Thornton Tomasetti to a national company, besides getting enough growth to pay out the original stockholders, was to grow the ACE Program. This would enable Thornton Tomasetti to become known as a national firm. So the first city where we expanded ACE was

Stamford, CT – because that is where I lived. The second ACE location was Newark, NJ.

Connecticut

In 1994, Ed Kasparek called me out of the blue mentioning something about talking to a New York-based MEP engineer who said that Thornton Tomasetti was looking to expand into Connecticut. Ed is probably the best business development person in the business. He's gracious, social, friendly and never gives up. He talked his way into coming to meet me in New York. I asked Larry Hine to join me. Ed showed up and explained that Malafronte and Kasparek (M & K), a Bridgeport-based MEP engineering firm, was interested in joining up with a national firm. To show the power of Ed's selling abilities, he sold us on the deal even though we had no idea or interest in restarting a MEP arm in New England or Connecticut.

So Larry and I went through the M&K finances and confirmed that they were indeed in financial trouble and decided not to do the deal. We briefed Richard Tomasetti and decided to invite Ed to the New York office to break the bad news that we would not be interested. Ed showed up to meet with Richard Tomasetti, Larry Hine, and me. Although we had decided to not do the deal, Ed started to do his thing and to our surprise Richard said yes. Ed is one of the most persuasive people that I have ever met and he does it so subtly and with such grace. So despite our misgivings, Larry Hine re-negotiated all of M&K's debt and we did the deal. It was never a very good deal, but I got what I wanted – Ed Kasparek.

Dallas

One of my classmates at Manhattan College and NYU was Dr. Leo Galletta. Unlike me, who dropped out of ROTC after two years because

I couldn't stand it, Leo stayed in for four years and was commissioned as a lieutenant in the Air Force. He spent a wonderful time in a Minuteman silo near Cheyenne, WY. After getting out of the Air Force, Leo actually came to NYU to get his PhD and we rekindled our friendship. At one time, he was a full-time employee of Thornton Tomasetti. At another time, he was a full professor at Cooper Union. Actually, Leo was the first person to introduce Richard Tomasetti and me to Hunan and Szechuan Chinese food. Until about 1970, the only thing we could get in a Chinese restaurant was Chop Suey and Chow Mein. Leo worked with me on the Valley Curtain and several other projects. He is a great engineer, a great salesman, a great eater, and a really nice man. His wife Mady is even nicer. Sorry Leo.

One day, Leo walked into the office and said, "I'm moving to Texas. I want to play more golf." Reluctantly, I accepted his resignation and said, "You'll be back." In 1994, I had a Building Seismic Safety Council (BSSC) meeting in North Dallas, so I called Leo and asked where his office was so that we could get together. It turned out that his office was right across the LBJ Freeway from where I was staying. I walked across and we had dinner. Over dinner, I asked him what was happening at his company. He responded "nothing." In fact, his CEO, was firing the wrong people, angering all their clients, and wasn't interested in spending money on technology. So I told Leo to have him call me and I would see what I could do.

Several weeks later, I called the CEO and flew to Dallas to inquire about an acquisition. Ellison Tanner had been the largest structural engineering firm in Texas with offices in Houston, Dallas, and Austin. They had designed many more high-rises than Thornton Tomasetti. Elmer Ellison was a heavy smoker and had died quite young. Tanner was not a great manager and had lost some of the superstars who worked for Ellison Tanner in the Houston office, including Joe Colaco.

After a short negotiation, we did a deal with the company similar to the deal we did with Eli Cohen. We closed the deal and within six months, we had won the Fort Worth Museum of Modern Art (MOMA) with a fee equal to the Dallas office's annual revenues. On this project we were able to work with a very famous Japanese architect, Tadao Ando, and produce a masterpiece.

What we really wanted out of the deal was Leo and the key staff and we got this. The Dallas office became one of the most successful offices within Thornton Tomasetti. It still exists today and Leo Galletta works there part-time. The Dallas, New York, and Kansas City offices just won the Minnesota Vikings football stadium in Minneapolis. For a while, Patricia Coleman ran the Dallas office and is very involved today with ACE Dallas and ACE National and the ACE National Affiliates Council (NAC).

California

In 1990, when we won the Anaheim Mighty Ducks Arena, I asked around as to who would be a good local small structural engineer to work with. The City of Anaheim mentioned Coil and Welsh. I called John Coil and asked him if he would collaborate with us on the project. He instantly said yes. John Coil was and still is a great man. Shortly after meeting John, he called me and asked me if I would be interested in joining the board of the Applied Technology Council (ATC). After explaining to me what ATC did, I said yes. This connection enabled us to start ACE in Santa Ana/Tustin, CA.

The eventual acquisition of Coil and Welsh propelled Thornton Tomasetti into playing on the national stage. Thornton Tomasetti's West Coast offices consist of Newport Beach, Los Angeles, San Francisco, Oakland, and San Diego – this is proof that the strategy worked. The really strong structural engineering firms were in San Francisco. The weaker structural engineering firms were in the Los Angeles area. My strategy was to come in under the radar screen into the soft underbelly of California and get Thornton Tomasetti established and go from there.

By serving on the ATC Board, I developed relationships with most of the top structural engineering firms in California. Most Los Angeles offices had satellite offices in Orange County, primarily to ease the commute for those people who wanted to live in Orange County, but didn't want to commute into Los Angeles.

The relationship with Coil and Welsh was great. Thornton Tomasetti was now bicoastal. The firm could now build its seismic engineering reputation by hiring the best and brightest out of Berkeley, Stanford, and the other great engineering schools on the West Coast. In recruiting in the East, we could offer the California experience – by recruiting in the West we could offer an East Coast experience. A lot of engineers go to California to get their master's degrees. Some stay and some come back to the East Coast. Being bicoastal is a big plus in maintaining a company staff and not losing staff because of a spouse getting relocated. Coil and Welch was not a big firm, so it was uncomfortable with large projects. We had to bide our time until we could grow the office and win large jobs.

Having Coil and Welsh helped us win the San Diego Padres Ballpark – Petco Park, the conversion of the Anaheim football stadium back to baseball only, and the crown jewel – PacBell Park in San Francisco. Through ATC, I met Mark Saunders and Bill Holmes of Rutherford and Chekene and they agreed to associate with us on PacBell Park, now known as AT&T Park, home of the San Francisco Giants.

I learned that most California-educated structural engineers are overeducated and live in fear of their professors showing up on a blue ribbon peer committee to review their work. The California structural engineers are all members of the Structural Engineers Association of California (SEAOC). The really good ones are also involved in ATC. The upgrade of the National Earthquake Hazard Reduction Program (NEHRP) codes, sponsored by FEMA, were managed by ATC and the Building Seismic Safety Council (BSSC). As a result of this code involvement, the really good structural engineers in California knew exactly what was happening in the next code cycle. So whenever a blue ribbon committee showed up on a project and you were designing to the old code, a former professor could say, "Shame on you – you know the codes are changing, why are you using the old code." Engineering firms, however, from the East, from Chicago, from Seattle or Texas really don't care about the new code because by law, they only have to design for the old (existing) code.

If you look at the history of tall buildings in San Francisco and Los Angeles and overall design by out-of-state structural engineers, in most cases, developers do not really want to spend more money designing to a code that is not yet enacted. Most of the high-rise buildings in San Diego are designed by Canadian engineering firms from Vancouver for the same reason.

It was no accident, then, that all these sports facilities were done by Thornton Tomasetti – we were really not a California-based firm and even though we may have designed somewhat to the new codes, it was less of an issue on sports facilities because they are such large public venues. So Thornton Tomasetti was recognized as a very strong structural engineering firm with strong seismic engineering skills, but we would design buildings to a reasonable standard that was in effect at the time of the design.

Meeting John Coil and being asked to serve on the ATC board led to serving on the Building Seismic Safety Council (BSSC), the FEMA Multi-Hazard Mitigation Council, and ultimately serving as the chairman of the National Institute of Building Sciences (NIBS). At about the same time as I developed a relationship with Henry Michel, I learned about the Civil Engineering Research Foundation (CERF). CERF was located in Washington, DC, and run by Harvey Bernstein, a structural engineer with degrees from the New Jersey Institute of Technology and Princeton. Harvey started his early career working in deep submersibles also known as submarines and then segued into association work in the Washington area. Harvey is one of the few out-of-the-box thinkers in Washington. Harvey joined McGraw-Hill Construction as head of analytics. Harvey does a terrific job. He has a great family and is a great friend.

Another great friend, Norbert Young of McGraw-Hill Construction, was born in Maine and attended Bowdoin College and the University of Pennsylvania for architecture. He started his career with an architectural firm Bower Lewis and Thrower, then joined Scott Toombs and developed projects in Princeton, New Jersey and other places. I first ran into him when he joined Lehrer McGovern Bovis (LMB) as a person in charge of special projects and business development. Norbert was instrumental in winning the construction program management for the Atlanta Olympics and the Atlanta Braves baseball stadium. We became very good friends and worked together on many projects. Norbert left LMB to become president of McGraw-Hill Construction, which publishes "Engineering News Record" and "Architectural Record," along with many other great magazines. He put "ENR" and McGraw-Hill Construction on the map.

Norbert and I were very active at the National Institute of Building Sciences (NIBS), which was the home of the International Alliance for Interoperability. As I moved up the chain at BSSC, CERF, and NIBS, Norbert was an active member on most of the projects and issues. We both became frustrated with the bureaucracy at NIBS and left, along with Patrick MacLeamy, Chairman of HOK.

Norbert became a strong advocate for ACE. He and his wife Christine became very good friends with Carolyn and me, travelling with us on two occasions to France and Turkey. The first trip was a river cruise for just the two couples on a 40-foot boat with bicycles for seven days on the Canal du Nivernais in France. Carolyn and I went to Normandy to see the D-Day sites and then drove back to Paris to meet Norbert and Christine at the train station and took the train for one hour to the embarkation location. It was a fabulous trip, the weather was great and the effort to drive a 40-foot boat with a maximum speed of five knots was pretty easy. We cycled into towns every morning to get our baguettes and butter. A couple of years later, Norbert and Christine joined us in Istanbul for three nights, after which we flew to Gocek, Turkey and sailed a 49-foot Bavaria sloop on the Lycian coast with a captain and a cook.

Stringing together these elegant solutions, and working the network I developed, led to the success of Thornton Tomasetti in many different ways.

13
Oklahoma City

IN 1995, I WAS FLYING into New York City when I heard about the Oklahoma City bombing. It was hard to fathom that a truck bomb in one truck could take out 60 percent of the Alfred P. Murrah Federal Office Building. At the time, I was actively involved with FEMA and National Institute of Building Sciences (NIBS) Multi-hazard Mitigation Council, Building Seismic Safety Council (BSSC), and The Applied Technology Council (ATC). When something like Oklahoma City occurs, FEMA has search and rescue teams that they deploy to help in the rescue operation. These groups are multidisciplinary teams of architects and engineers, heavy equipment operators, search and rescue K-9 teams and construction workers. In addition, FEMA generally dispatches a Building Performance Assessment Team (BPAT).

I was selected along with Gene Corley, Mete Sozen, and Paul Mlakar to deploy to Oklahoma City to observe the Murrah Building collapse about 14 days after the event. Just to confirm how screwed up the federal government is, the GSA District which controlled Oklahoma City was based in Dallas. Although the GSA in Washington and FEMA had authorized our visit, the local GSA people would not allow us access to the actual collapse site.

Photographic documentation of the event was quite prolific, so our team really didn't need access to the site to determine what happened. All we needed to see were the original drawings and blast input from Paul Mlakar. We worked for a couple of weeks, prepared a report and within two months concluded that the original design in the mid-70s was actually almost perfectly done and the drawings were very complete. The only problem was that since seismic design was not part of the code for Oklahoma in the 1970s, the time of the design, all of the lateral load resistance for wind and earthquake was located on the far side of the building away from the street where the truck was parked. This was a perfectly acceptable design for the time, but resulted in the street side of the structure having little or no resistance to progressive collapse.

We gave our opinion that had the building been designed for seismic forces, 60 percent of the building would not have collapsed. We advocated that all buildings, with the potential to be attacked by terrorists or even hit by a big truck, should be designed for minimum seismic forces, as well as disproportionate collapse and progressive collapse elimination. Another way of describing these techniques is that if you knock out one column, only that portion supported by the column should come down and collapse. The terrorists in Oklahoma City knocked out one column and took down 60 percent of the building. Since the Oklahoma City event, all government buildings are designed for disproportionate collapse and progressive collapse.

14
Thornton Tomasetti Ownership Transition Plan

LEADING UP TO 1995, I had prepared the top management of Thornton Tomasetti to look forward to a transition to the next generation of owners and leaders. Rumors were already spreading that some of the next generation were about to be poached, recruited, or pirated by some competitors. There was no way that I was going to allow procrastination and punting to lead to losing some of the superstars in our firm. Some of my colleagues were not necessarily as anxious to preserve the next generation of management, as I was. They suffered from the same thing that most firms suffer from. They think because they are the boss and the King, they will always be the boss and the King. They also are in denial about getting old. This is what leads to the demise of almost every engineering firm in the United States that waits too long to start the transition.

The next generation of engineers was on the average about forty-years-old and had been with the firm 15 to 20 years. These are the people you can't afford to lose. Thornton Tomasetti was owned by me and Richard Tomasetti, at a combined 68 percent of the stock with four other minor stockholders: Joseph Zuliani, Abraham Gutman, Jagdish Prasad, and Dan Cuoco at a combined 32 percent of the stock. Joe and Jay were older than me, Abe was the same age as me, and Dan was about 10 years younger than me. Richard was two years younger than me.

Having lived through the acquisition of the company by Gable Industries, which was actually quite successful for us, I didn't want to go through another sellout of the company to a larger firm. Many of my older colleagues in other firms who sold companies to larger firms admitted to me that during the 10 years prior to retirement, they worked harder than they had worked in their whole life to prove to the new company that what they paid for it was worth it. There was no way I was going to ever work for anybody again. As a result, whether Richard liked it or not, we were proceeding with our plan.

We hired Zweig White. Richard Tomasetti and Larry Hine, working with Ray Kogan and me, came up with a transition plan, which is explained in my "15 Steps to Success" located later in this section. All of the key staff knew exactly what my intentions were. They all knew that I wanted to be out in 10 years and Richard wanted to be out in 12 years and that during the first 5 years, they would end up gaining control of the company. Just based upon the demographics of the six of us, it was obvious that under my plan over 50 percent of the stock would be available to the new owners within five years. The next generation, led by Tom Scarangello, Robert DeScenza, Dennis Poon, Aine Brazil, Manny Velivasakis, Joel Weinstein and Steve Dennis, knew exactly where I wanted to go. Some of the present owners wanted to delay the process – more punting and procrastination. I live by the philosophy, "Make a decision and go for it" or "Just do it!" We trusted these people and had worked with them for 15 to 20 years. Most of them were hired out of classes taught by either Richard or me.

In 1996, a total of 12 employees were given the opportunity to buy shares of the company's stock. Several were entitled to a one-time buy and others were entitled to multiple-year buys. The initial group was comprised of Aine Brazil, Steve Dennis, Udom Hungesbruke, Len Joseph, Paul Lew,

Dan Margulies, Robert Nacheman, Dennis Poon, Tom Scarangello, Manny Velivasakis, Joel Weinstein, and Leonid Zborovsky. Of these stockholders, six managing principals were given the opportunity to make multiple-year purchases and have more comprehensive agreements with expanded rights. As time evolved, Bob DeScenza, Joe Burns, and Daniel Marquardt in Chicago were also given opportunities for multiple-year stock purchases. By 1997, Leo Galletta in Dallas, and Frank Gallo and Mike Makara in Trumbell, CT, also became stockholders.

By May 2000, the company had 31 stockholders, including new stockholders Narinder Chhabra, Chris Christoforou, Matt Hoelzli, Anjana Kadakia, and Ed Kasparek. Richard and I were a great team. During the transition, Richard took the lead and did a great job in documenting the ownership transition, the financial plan, and the compensation system. We believed in opportunity, just as Lev Zetlin had provided to us. We wanted to pass the baton of leadership to the next generation of leaders. Looking back on how we negotiated our exit from ownership, it was the right thing to do.

Philadelphia

In the late 90s, we acquired a small office near Philadelphia in Claymont, DE. We then moved it into Center City Philadelphia so that we could be downtown. On one of my trips, I left the office and was walking down Sansome Street toward the Society Hill Sheraton, which we had designed for Bill Rouse and Joe Denny with Brennan Beer Gorman, and I saw these three young African-American women in khakis and Oxford cloth blue shirts. I asked them what organization they were affiliated with. They said the Charter High School for Architecture and Design (CHAD). I asked them to tell me a little bit about it. They told me to go in and speak to the principal.

I walked in and introduced myself and within an hour I was on the board of CHAD. CHAD was a natural vehicle for ACE to associate with in order to get ACE of Eastern Pennsylvania going. I stayed on the CHAD Board for about four years. As we started to launch the Philadelphia office, I got to know Tony Naccarato of O'Donnell-Naccarato Structural Engineers. I asked him if he was interested in participating in ACE and he said yes. So, after 10 years, my daughter Diana and Tony basically ran ACE of Eastern Pennsylvania. It was very successful with a large number of students, numerous scholarships, and a great scholarship breakfast, which continues today and is generally sold out. About two years ago Diana stepped down from Eastern Pennsylvania to join ACE National to be the Northeast regional coordinator.

15
Maryland's Eastern Shore

CAROLYN AND I HAD STARTED looking for an exit strategy to bail out of the high-priced New York area. Although we lived on the water in Stamford, CT, in a very nice house, we were on a quarter acre parcel with six feet on either side of the house to the property line. This meant that there was only 12 feet between houses. I wanted more room. From 1990 thru 1994, we started looking at coastal locations from South Florida to Amelia Island near Jacksonville, FL, to Georgia to South Carolina.

One day, Carolyn reflected on two wonderful sailing excursions we experienced on the Chesapeake in June 1983 and June 1984. The first charter was at Great Oak Landing on the Upper Chesapeake Bay. On that cruise, we visited Easton, St. Michael's, Oxford, Annapolis, MD, and points north. The second trip we chartered two Dickerson 37-foot boats out of La Trappe Creek and sailed Cambridge, Oxford, and all the way down to Crisfield, MD. During this cruise, we were on the Choptank River and we realized that a very strong thunderstorm was blowing over Virginia from the west. We motored into La Trappe Creek and anchored in Sawmill Cove, a lovely spot. As we were motoring in, we noticed a kid on a sail board who was oblivious to the storm and was in big danger. Since we were towing our 13-foot Boston Whaler, Charlie III went out and towed him in before the storm hit.

What a spectacular storm. Fortunately, the anchors held. As the sky cleared and the sun came out, we noted the entrance to La Trappe Creek – one of the most glorious places we had ever seen.

So when it came time to look for real estate, Carolyn said, let's look at Trappe, Easton, St. Michael's, and Oxford, MD. So in the fall of 1997, we started our search. Talbot County, MD has 780 miles waterfront property. It is a county on the Chesapeake Bay with more waterfront lineal mileage than any other county in the United States.

After establishing 2000 as the pivotal year in my exit strategy, we decided that we would buy land on Maryland's Eastern Shore.

Easton, MD, is centrally located between Washington, DC, Baltimore, MD, Wilmington, DE, and Philadelphia, PA, which is only 120 miles away, and it was only about 200 miles to New York City. All the other places we were looking at required getting on a plane and generally changing planes somewhere in order to get back to the Northeast. This did not interest me.

16
Washington, DC Office

ABOUT THIS TIME I WAS on the ATC board and I got to know a Washington structural engineer named James Cagley. Also about this time Mark Tamaro, with Thornton Tomasetti, came to me in the New York office and indicated that, in light of the fact that he had proposed to his Lehigh University sweetheart and she had a job in Washington, DC, he was moving to the Washington area and joining Jim Cagley and Cagley Associates. Mark stayed with Cagley a short time, but nepotism got in the way and he ended up with James Madison Cutts. One day, I asked Jim Cagley to tell me about James Madison Cutts. Cagley said everybody has tried to acquire Cutts, but no one will ever succeed. Under my breath I said, "Just watch." So I called up Mark and asked him to introduce me to Jim Cutts, which he did, and within six months Larry Hine and I had closed the deal to purchase James Madison Cutts, a 12 to 14 person highly-rated structural engineering firm on 2000 L Street, NW, in Washington, DC.

This was all part of my strategic plan to move in 2000. When Becky finished 10 years at Greenwich Academy in Connecticut, we would enroll her in Gunston Day School, the only excellent co-ed non-secular, private high school on Maryland's Eastern Shore. At this point, I acquired an office in Washington, DC, because no one else in the New York office wanted to move. A DC office would enable us to acquire more federal government work, which really helps out in recessionary times. Who knew that the recessions were coming? So, with five years to go in my ten-year sellout, Carolyn and I purchased 17 acres with 3000 feet of shoreline on the Miles River. The lot had deep water – 8 feet to be exact – so that I could keep my beloved Elegant Solution III at the dock on our property.

Upon returning home, after purchasing the property in January 1998, I called my good friend and architect, Alfredo DeVido and told him I wanted to live in a DeVido home. Between 1998 and 1999, we collaborated with Al on the architecture and engineering of our 7,000 square-foot house with 300 windows. He was the architect of The Filene Center at Wolf Trap Farm in McLean, VA, and Mann Music Center (originally Robinhood Dell West) in Fairmont Park, Philadelphia, PA, the summer home of the Philadelphia Orchestra. Thornton Tomasetti and I were the structural engineers on each of these projects, which used a lot of wood and sat over 5000 people – something that fascinated me. I also worked with DeVido on many large waterfront homes in the Hamptons and Connecticut. All of DeVido's designs for the theaters and most of his residential designs relied on exposed structure and its interaction with the architecture. In addition, he had a knack for bringing in complex contemporary homes at unbelievably reasonable construction costs. He created a masterpiece. Our house is featured in an Australian publication, *100 More of the World's Greatest Houses.*

In 1993 I co-authored a book, *Exposed Structure in Building Design*, with Richard Tomasetti, Janice Tuchman, and Leonard Joseph. I picked prominent architects, including DeVido, Cesar Pelli, and Gyo Obata, as well as a number of structural engineers, to be interviewed about why they used or didn't use exposed structure in their buildings. By publishing the book, we hoped to promote Thornton Tomasetti structural engineering expertise on exposed structure projects.

Later, in 2006, I was awarded Honorary Membership in the American Institute of Architects, one of the highest honors a structural engineer can receive. Many of the letters of recommendation for this honor, written by prominent architects, commented on my innovation as an engineer as I partnered with some of the country's strongest architects to design some of the most beautiful structures around the country. These collaborations were an integration of great architecture and great design – Elegant Solutions that were both aesthetically pleasing and technically cool.

In August of 2000, we relocated to the Eastern Shore of Maryland and moved into the house in November of 2000, just in time for Thanksgiving. All the brokers said you could play golf all winter in shirtsleeves in Talbot County. So I bought kayaks and canoes and wouldn't you know it, the Miles River was frozen solid when Carolyn's entire California-based family showed up for Thanksgiving – so much for golfing in shirtsleeves all winter.

We also closed the deal with James Madison Cutts in 2000. We moved Jon Kopp from Chicago to help run the office and we picked up some superstar engineers in Wayne Stocks and Mark Tamaro. From 2000 to 2005, I basically had a four-day weekend. I drove to our DC office on Tuesday at noon and drove home on Thursdays at noon. It really was a premature retirement and I realized that I didn't really like it. What I did like, however, was the personal renaissance that I experienced during this period of my life. I was now surrounded by a bunch of young engineers who had never experienced my style of management. The owner of the firm, Jim Cutts, had lost his future management team every five years to competitors in Washington, DC, because he was unable to keep these young engineers motivated. I was able to rejuvenate the office – bringing in the DC Nationals Ballpark and the million square-foot jewel at Johns Hopkins Hospital in Baltimore just before I stepped down.

Although I had a very good relationship with the key people in New York, they had stopped listening to me to some degree – I think it was because they had known me for so long. Now, I had all these young engineers, bright lights who wanted to experience the Thornton Tomasetti growth, the business development machine, and the high external profile from publishing their work. With Wayne, Mark and Jon, as well as Zach Kates, I experienced working again with engineers with great ambition and hope for the future. It was fabulous.

17
Music

AS A CHILD, I LOVED to go up into the unheated attic in our house and listen to our boxed set of Strauss waltzes, including "The Skater's Waltz" and "The Artist's Life," on my parents' windup RCA wooden Victrola. When I wasn't in the attic, I spent a lot of time in the kitchen playing on the floor while my mother cooked. She always had the radio on and enjoyed Broadway show tunes by Rogers and Hammerstein, Rogers and Hart, and Harold Arlen. When she would sing these tunes, she was always happy. When "South Pacific," "Carousel" and "Oklahoma!" came out on Broadway, the music was always playing at the nearby Shore Haven Beach Club where we spent a great deal of time. My mother also loved Cole Porter and Gershwin. Through her, I developed a profound appreciation for the music.

In 1955 when I was 15, Sun Records released recordings of Elvis Presley, Carl Perkins, Roy Orbison, Johnny Cash, and Jerry Lee Lewis. In 1955, Bill Haley and the Comets came out with "Rock Around the Clock" and The Platters came out with the "Great Pretender" and "Smoke Gets in Your Eyes." As Mississippi Delta music moved up the Mississippi River, we started listening to Alan Freed from Cleveland and Chess Records in Chicago introduced Bo Didley, Chuck Berry and Little Richard. We never listened to the lyrics. Later in life, I realized that the lyrics were pretty foul. No wonder our parents thought we were going to the dogs.

When my brother Bill was a freshman in Manhattan College, he betrayed us and brought home "Pictures at an Exhibition" by Russian composer Modest Mussorgsky; "The New World Symphony" by Antonín Dvořák; as well as works by Nikolai Rimsky-Korsakov and Tchaikovsky. I couldn't believe that he was betraying our loyalty to rock 'n roll. I listened to each of these four LPs about five times and loved them.

In the early 60s as I developed a great interest in classical music, I started attending the New York City Ballet, the American Ballet Theatre, New York Philharmonic, and many other music venues at Lincoln Center with Patricia. After I got my PhD in February 1961, I was so fed up with technical education that I went to NYU's continuing education and took a course called "Great Symphonies and Concertos" with David Randolph, the director of the Masterworks Chorus. He took us through Beethoven's Seventh Symphony, which is a masterpiece. He explained the promenade sequence, introducing each instrument, and I never forgot it.

I attended a few chamber music concerts when I lived in Westchester County, but it never really caught on until we moved to the Eastern Shore of Maryland. Carolyn had been a concert mistress violinist in her middle school orchestra in Santa Monica, CA, and always liked chamber music. So in 2000, when we moved to Easton, Carolyn played tennis with several board members of Chesapeake Chamber Music. She later joined the board and we have participated in the annual Chesapeake Chamber Music Festival, Gala, and Competition every year since then.

During this time, we also used our Eastern Shore home to sponsor an annual weekend music event, the ACE Celebration Party, for the benefactors of the ACE Mentor Program. The first evening included cocktails and hors d'oeuvres, a chamber music concert, and a five-course dinner. The next day included a boat cruise of the Miles River near our home and a buffet brunch. This was an excellent event that helped to develop longstanding relationships with the program's sponsors.

18
The Fall of the World Trade Center

I WAS IN NEW YORK having breakfast at the New York Yacht Club on 9-11. Breakfast was with Ray McGuire, one of the top labor lawyers in the New York construction industry, who represents Contractors' Association of Greater New York (CAGNY) and tends to manage the relationship between the big general contractors in New York and the organized labor unions. The unthinkable happened at 8:46 a.m., when a yacht club employee walked into the dining room and said that someone flew a plane into the World Trade Center. Ray looked at me, I looked at Ray and we shrugged it off, it must have been a Cessna or a Piper Cub. Well, 15 minutes later, the same person ran in and said that someone flew a second plane into the other tower. We looked at each other and said, "It's time to get out of here." I walked to Sixth Avenue and 44th Street and looked south and saw the World Trade Center in flames, framed on the right by my beloved World Financial Center at Battery Park, which my firm had designed.

Being a New Yorker by birth and knowing the subway system, I jumped on the F train at 42nd Street, went two stops, got off at 23rd Street and walked into my office building at 20th and 6th Streets. The place was different. Women, secretaries, receptionists, and staff were crying. They all had friends in the buildings; none had any idea whether or not their loved ones were okay.

Being in NY on this day was one of my infrequent visits to the city as I had relocated to Maryland's Eastern Shore about a year earlier and was working predominately in our Washington, DC office. As I walked around the NY office, the atmosphere was different, people I knew were not working, they were in our conference rooms and their eyes were glued to TV sets. I proceeded up to our upper floor, the eighth floor, and went out on the roof. I watched the fires in the towers. From this vantage point, my colleagues and I watched the tops of the buildings become engulfed in flames and smoke. The 20 engineers and architects and people standing around me, all of whom had firsthand knowledge of fires in steel high rise office buildings, knew one thing for certain – the towers weren't going to collapse. But then, the unthinkable happened!

As time went by and I started to study the structure in much more detail than I had ever studied it before, with the help of Len Joseph, one of the brightest people at the Thornton Tomasetti Group, I uncovered a lot of information about the WTC floor system, the WTC structure, how the WTC building resisted wind, and how the WTC was put together. The more I learned about the WTC, the more concerned, the more upset, and the more outraged I became about why it collapsed. There were problems that should have been known, but that is the subject of another book.

19
Travels with Charlie (and Becky)

SOMETIME IN 2004, WE BOUGHT Becky a new Honda Civic, following her graduation from Gunston Day School in Centreville, MD. On June 26, 2004, we were hosting Chesapeake Chamber Music's Annual Angels Concert at our home. At about 3:30 p.m., as the 100 guests were arriving and the musicians were practicing, I heard the phone ringing. The music had dampened out the ability to hear the phone, which had rung several times.

I learned later that Becky had called a couple times and left several messages. When I finally picked up the phone, Becky said, "Hi Dad, don't be alarmed, but I am in an ambulance. We totaled the car. I was thrown out the window, but all I have is bruises and road rash." Rather than alarm Carolyn, I just said to Don Buxton, the executive director of Chesapeake Chamber Music, that I had to run into town for a quick visit. I arrived at the hospital figuring that they would keep Becky overnight. But they said she was going to be released in about a half-hour. I told them I would be back to get her and I drove home and caught three of the five pieces played by great chamber musicians. By the fourth child you become more relaxed about these types of things.

Unbeknownst to me, Carolyn had listened to the phone messages and was well aware of exactly what was going on. My little ruse had been exposed. Just before the last chamber music piece, I jumped back in the car and went and picked up Becky and brought her in the side door of the house as if to "smuggle" her in. However, Tara Helen O'Connor, one of the world's greatest flute players, was aware of what was going on, and she came upstairs and sat with Becky for quite a time to chat with her.

The accident had occurred when Becky allowed an underage girl, without a driver's license, to drive the car. Becky was in the front passenger seat without a seatbelt changing into a bathing suit with the windows open. When the girl over steered then overcompensated, the car hit a ditch head on and rolled over, throwing Becky out the window. Becky was unconscious and taken to the local hospital. Although others had minor injuries, they are all lucky to be alive today.

Just prior to June 26, some of Becky's "friends" had dumped our planters with a lightweight Pearlite-Premix soil into the pool, clogging everything. We called the pool company and they came quickly and remedied the situation, but the handwriting was on the wall that I had to get Becky out of town.

I had spent quite a bit of money on new marine electronics for Elegant Solution III and was planning to cruise to Maine from the Chesapeake. Upon second thought, I concluded that going away and leaving Becky and Carolyn alone was not a great idea. So I told Carolyn I was taking Becky on a road trip and not bringing her back until she agreed not to hang out with the friends she had. On the trip we traveled through West Virginia and ended up in Covington, WV, near Warm Springs. We then went to Lexington, KY, and on to Paducah, KY, the quilting capital of the world. That evening, we were in a Japanese Steak House eating dinner. The other people at the table kept eying Becky suspiciously as she had two black eyes from the car accident. As I realized what was going on, I introduced Becky and me to the group at our table and explained about her accident a few days before. We got along grandly after that.

We then ended up in Branson, MO, which Becky called "the Las Vegas for older folks," because of the buses carrying senior citizens. In Dodge City, KS, our hotel was downwind from a cattle feed lot – a great place to spend the night. We then stopped in Denver for a few days to see Charlie, attending a Rockies baseball game there. Carolyn flew out to join us for a few days. Becky and I resumed the trip for several more days, driving through the Grand Tetons in Wyoming to Yellowstone, on to Mount Rushmore, the Badlands, and then back home. This was the first of several great road trips we did together. In 2005, we went to the East End of Puerto Rico, including the Island, Puerto Rico.

On another trip, we had read about Merritt Island, a bird sanctuary near Cape Canaveral, so we flew to Orlando, rented a car and drove over to Cocoa Beach. We couldn't find a bird. We had talked to a friend and colleague, Charlene Whitney Edwards, about Wakulla Springs near Tallahassee, FL. So we turned around and drove across Florida to Tallahassee to see the springs. It was an awesome experience. This was a large sinkhole where they filmed such movies as "Tarzan" and "The Creature from the Black Lagoon" in the 50s. I had never seen anything like it - 500,000 gallons of water a minute comes out of this hole in the ground and it starts a river. The water is crystal clear. The alligators, turtles, fish, and birds all coexist in this unfenced zoo. There is a boat tour of the river, which includes seeing a skeleton of a Mammoth, if the water is clear enough.

We got there late in the evening and were able to get the last room in Edward Ball's mansion. Ball and duPont controlled most of Florida's timberland and donated the large estate to the State of Florida. We were there just in time to catch dinner, which was rather spartan but good. We signed up for the morning trip on the tourist boat and we loved every minute of it.

We then proceeded down to Cedar Key on the West Coast of Florida right by the Crystal River. Cedar Key is a small waterfront town with galleries and restaurants and motels. While on the Crystal River, you can see the nuclear plant cooling towers in the distance. It's pretty much in the middle of nowhere. We then drove to Orlando and flew home.

We also did a family trip with Carolyn's sisters and family to the Galapagos Islands on a Lindblad Expeditions/National Geographic cruise ship, which was absolutely spectacular. Several years later, Becky and I did a Lindblad/National Geographic cruise to the Southwestern Passage of Alaska, from Juneau to Sitka, which was also an unbelievable trip. I kayaked and we hiked every day. We even went out in the Zodiak inflatables and saw humpback whales, an Orca killer whale, Steller sea lions, sea otters, grizzly bears, and mountain goats. This is a unique place in the world because you are at sea level and are looking at 14,000- foot peaks. In Colorado, you are at 7,000 feet looking at 14,000-foot peaks.

More recently, in July of 2011, Becky and I flew to Fresno, CA, and rented a car and visited Yosemite. We then left through the East entrance of Yosemite in the mountains and drove into Nevada and up into Idaho, visiting Idaho Falls, Sun Valley, and the beautiful green irrigated valleys. We then continued over to Yellowstone for our second visit. These were trips that the two of us wanted to do, but that Carolyn didn't necessarily want to do. As Becky developed into a pretty fabulous photographer, we purchased really good lenses and equipment for her and coordinated road trips which would capitalize on her nature photography. On this trip, Becky took unbelievable pictures of elk, moose, coyotes, waterfalls, and geysers. She even got chastised by a park ranger for getting too close to a full-grown bull moose, which I thought was pretty funny.

Becky said later that these were bonding trips for her where she learned a lot about life. Through learning about my experiences in life, she learned about how I became the man I am. She said she learned everything from financial responsibility to personal responsibility. Being trapped in the car with me for hours at a time on these trips, she learned things even if she didn't want to. I felt that every trip I took with Becky was a concentrated time with just her. This was a unique experience for her as she had no siblings to compete with.

It was a concentrated interaction for the two of us, in limited space and time, between breakfast and bedtime. In addition to intellectual conversations, I was able to convey everything my father taught me – common sense things about ethics and honesty. She couldn't leave when the conversation got uncomfortable. We also connected in outdoor activities, which I enjoyed. Her love of photography also created a bond between us.

20

Establishing Charles H. Thornton & Company and 15 Steps to Success

MY DEPARTURE DATE FROM THORNTON Tomasetti was scheduled for December 2005 – the end of the 10-year transition. As a strategic planner and strategic thinker, I realized my life was to have a plan as well. In May 2004, 19 months before my departure, I sent two of my colleagues at Thornton Tomasetti an email to propose what my role would be once I ceased to be a chairman and a stockholder. In May 2005, I was frustrated with the progress we were making – more punting and procrastinating. While sitting in my hot tub at my new house, drinking really good red wine, I realized I had the making of a management team within my own family. I thought to myself, why not name the company, Charles H. Thornton and Company to bear my family name?

Charles H. Thornton and Company, LLC, was formed to provide strategic and management consulting; human resources and employee benefits consulting; financial consulting; investigative and expert witness services; structural concept development; and structural steel consulting services.

In the process of forming Charles H. Thornton & Company, I developed "Fifteen Steps to Success." As the owner, or major owner of a company, have you ever wondered why one of your competitors appears to be more successful; how to keep the momentum going that has made your firm successful; or how to insure the management of your firm continues beyond your involvement? These questions are not answered because of luck, but instead due to continued planning – defining a mission, creating the vision on how to attain it, and setting goals.

The process, the implementation, the review, and the team building require a company-wide commitment, but with this commitment, most, if not all of, your company goals, vision, mission and success will be accomplished. There is no cookie-cutter approach since each company's culture and objectives vary, but the process remains constant. The Fifteen Steps to Success are:

1. Interest, Will, Desire, Commitment to the Process:

It is easy and often too late to sit down and say what if we did that; we wish we had done that; that made sense why didn't we see it coming. Nothing happens unless you are committed and have the will, interest, and desire to make it happen. Many people talk the game but few take the formal steps to make it happen. You and the rest of the management team must sit down and determine that you are all committed and all have the interest, will and desire to implement this plan.

2. The Mission:

Where do you want to be in three, five, ten, fifteen years from now? How do you wish to be viewed by your clients, your staff, your peers and most importantly yourselves?

3. The Vision:

The vision is how you plan to carry out the mission statement. Successful businesses, work on short-term, mid-term and long-term business plans, as well as strategic plans and defining and implementing the plans to minimize business risks. Five-year plans are necessary to achieve significant growth, as well as management and ownership transition.

Depending upon the age of the existing owners and stockholders and the age of the new proposed owners and stockholders, a ten- or fifteen-year plan may be essential. The will, the interest, and the desire are the foundation of a successful plan. If you truly want your company to grow and you truly want to transition the management and the ownership to the next generation, you must have a plan. You must know about it, the staff must know about it. In most cases, plans are kept in people's heads, they are not communicated throughout the company and they never happen.

4. The 3, 5, 10, or 15-Year Plans:

Financial planning is essential to the process. The exercise in the beginning is as important as the results. It forces management to examine the company and plan for the necessary growth and profitability to insure a management and ownership transition, proper cash flow, merger and acquisition opportunities, staff acquisition and retention and business and client opportunities. It takes the commitment of the management team and key personnel to sit down area by area and examine all the alternatives to come up with the optimum plan.

The process is not "pie in the sky;" it is not an arbitrary percentage growth over the previous year; and it is not the use of a crystal ball. The process of meaningful plans is an examination of how the company does business, would like to do business, and examines the prudent business risks of cash flow, staff retention, staff motivation, mergers and acquisitions, marketing and business alliances, banking relationships, business areas and client mix within the company's mission statement and vision. It is a "living" plan that must be constantly reviewed and updated. A blueprint which if met will permit the defined goals and objectives to be met.

5. Compensation and Benefits:

Most people work for three reasons – enjoyment, satisfaction, and compensation. Each individual has varying degrees of the relative importance of each. The financial plan and management transition and ownership transition plan must combine a meaningful compensation and benefits package, as well as insuring a company culture and environment that enables key staff to find satisfaction, enjoyment and fulfillment in their work.

Management theorists have stated that a well-run company is one where the "individual's and company's objectives are one" and where the individual is "committed by involvement" (in the process).

6. The Marketing Plan:

What are you today and what do you want to be in five years? What is your scope of services today and what should they be in five years? What are your niche markets? What are you good at? What are your specialties? How do you differentiate your service from your competitors? Why will you succeed in market sectors, where others won't? A successful company has a marketing plan. Marketing plans are based upon underlying business within an industry as well as keeping one's finger on the pulse and monitoring new opportunities as global, national and local events, demographics, movements, change. Without a marketing plan growth and flexibility cannot occur at a sustained pace.

7. Staff Development Plan:

As the five, ten and fifteen year plans evolve, the compensation and benefit situation must be communicated to the staff. The future generation or generations needs to know what the plan is. They need to know what the company is going to do for them in terms of growth, job opportunities, expansion of roles and responsibilities, upward mobility, promotion schedule, and if and when they will become an owner, when and if they will become a principal, when they will become whatever they are going to become. Without a staff development plan companies will not grow.

Key personnel must be empowered to technically and managerially develop their staff – this may take outside assistance. It is important that all aspects of management be stressed – project management, the financial impact of decisions, the marketing, client handling and staff motivation.

8. Public Relations Campaign:

When the financial plans are developed, compensation and benefits plans are put into place, the marketing and staff development plans are in process, a coordinated, orderly public relations plan must now be worked on. The company must develop ways and means to make younger levels of management create public awareness within the press, the technical journals and within your major client's industries. This is a five-year plan as well.

9. The Financial Model for Growth:

In the process of developing the financial plan, management, top to bottom must also develop a financial model for growth. The projections for revenue, expenses, and profits must be put down in a precise plan. Financial growth models should take three paths: moderate, median and large growth patterns. In order to show that there is room for most of the aspiring managers to join the ranks of ownership, the company must grow in size, in revenues and at least maintain a profit margin capable of compensating the key staff and keeping them interested, committed and with the company. This plan would be developed in concert with top and middle management. Commitment to the plan by all parties is absolutely mandatory in order for it to work. It is the process by which buy/sell agreements and stock prices will be determined.

10. The Valuation of the Company:

Most companies reach out to certified public accountants or other financial managers to do an annual valuation of the company. A much more effective, less expensive and controllable way is to develop a formula based upon net assets (retained earnings) and a rolling average of profits. Between the prior section and this section, if you have three models for growth and the formula is flexible, then the value of the company, the value of the stock, the return on investment, the annual principal bonus and performance bonuses will all fall into place. Through this vehicle, the people selling the stock will know what they will be getting for their stock as they sell, and the people buying in will know what they will be paying. Since the plan and the formula is based upon three different models and the formula allows the value of the company and the value of the stock to go up and down, depending upon which growth pattern is ultimately achieved, this plan is totally flexible. Obviously, the better the company does, the more money goes into everyone's pocket. The better the company does, the higher the stock price becomes. The better the company does, the sellers net significantly more money in the sale of their stock. Although the buyers pay more, they also get a higher return on their investment. With this approach, everyone wins.

11. The Reasons to Buy:

A concerted internal public relations and awareness campaign must be launched in order to convince the younger generations that they want to be stockholders. Many times, younger employees are offered the opportunity to "buy a piece of the rock," but after they go home and discuss the situation with their spouse, they may decide that it is better to be invested in Microsoft, IBM, General Electric or whatever. Most technical people, especially engineers, architects, and construction people, tend to be "risk averse." Every day of their professional career they are involved in managing risk. Many younger employees are absolutely terrified at the thought of investing their hard-earned money in their own company's stock. However, what better stock to invest in, than one that is under your control. We all invest in the stock market and we have essentially no control. So a concerted effort must be launched to sensitize, inform and educate the key staff that the more members of the internal key staff that owns the company, the more successful the company will be. Again, it's "the intertwining of individual and corporate objectives."

12. The Formula:

There is no set formula for an ownership transition plan. Many individual considerations must be taken into account based on a specific company's culture, profitability, financial planning, and age of participants. However, what is constant is a mutual and cooperative interaction between the existing owner(s) and the perspective new owners/ stockholders. There must be a great deal of give and take between the seller(s), who think they're getting too little, and the buyers, who think the projections on which the formula is based, are unattainable. Frustration between the two sides is a guarantee, but with time and mutual trust and understanding the process will be successful.

13. The Stockholders Agreement (The Buy/Sell Agreement):

This step requires significant input from legal and accounting consultants. The basic formula and the basic timing; "the business deal;" the needs to be established by the principals, old and new. But the paperwork has to be put in place by experienced legal and financial consultants. Consideration must be given to questions of voluntary or involuntary separation from the company, life insurance, health insurance, and financing arrangements, defaults, and the like.

14. Benefits:

In any buy/sell deal, in any situation where people are spending hard earned money to buy into a company, there must be many, many, benefits to cover all of the contingencies and possibilities that can occur over the next five, ten and fifteen years. Life insurance, key manager insurance, cross sell insurance, disability insurance, all of these benefits must be in place to ensure that should tragedy hit a seller or a buyer that there is adequate insurance in place to pay off the buyer or the seller's estate in the event of death and/or disability.

15. Nepotism:

I realized after writing the 14 steps listed above, that I wanted to add a 15th step about nepotism, one of the main reasons why most organizations, companies, engineering, architectural, and law firms fail. Thornton Tomasetti instituted a very strong anti-nepotism rule and as a result no children of senior partners were hired in the company, except for part-time summer jobs. In my various conversations with industry leaders, nepotism is generally found to be the one factor that kills a company when the second generation takes over from the first and doesn't really deserve it.

If you take these 15 steps, manage them, implement them, maintain them and gauge your progress every year, your company should grow to the level you wish to attain.

PART THREE – RENAISSANCE/ DISRUPTIVE INNOVATION

21
Disruptive Innovation

"And it ought to be remembered there is nothing more difficult to take in hand, more perilous to conduct, or more uncertain in its success, than to take the lead in a new order of things. Because the innovator has for enemies all those who have done well under the old conditions, and lukewarm defenders in those who may do well under the new."
—NICOLO MACHIAVELLI, *THE PRINCE*

THE SUCCESS OF THE COMPANY, Thornton Tomasetti, was due to the relationship between its two leaders, Richard Tomasetti and me; our relationship with key staff; the staff's trust in us; and our trust in them. Equally important was our consistent innovation on all of our projects. Our plan was to transfer ownership to seven key people in the firm, making them as a group, majority stockowners within the first five years of a ten-year plan. This step was a leap of faith, but we knew the solid relationships between all of us would make the plan work. I never thought it was a risk because these seven people were absolute winners. Everybody won in this arrangement. By 2013, Thornton Tomasetti had 800 staff members in 27 offices around the world.

As the chairman and founding principal in a major structural engineering firm, I was constantly evaluating alternate structural systems for every project that we were contracted for to perform the engineering design. Until my separation from Thornton Tomasetti, it would have been impossible to own a proprietary company with a proprietary system. Who would think that we were being objective, if we were always advocating a system owned by us? The answer is nobody.

I waited until 2005 to form three new companies, Charles H. Thornton and Company LLC, TTG Corporation, and STRAAM, LTD. Even though I had succeeded in many projects with Thornton Tomasetti, I didn't want to repeat project types with my new companies that I had done while at Thornton Tomasetti. I wanted to innovate and do new things. I wanted to tread on new ground and to differentiate what I was doing from everybody else. I really wanted to practice disruptive innovation. I started out by branding all three companies using a talented graphic designer named Charlene Whitney Edwards of the firm Whitney Edwards, who lived in my new hometown, Easton, MD. She created award-wining marketing materials that really helped to differentiate me in the market.

Running a large company takes a considerable amount of effort. At the time I stepped down from Thornton Tomasetti it probably employed about 300 people; that number is now 800. Each of these employees has their own strengths and weaknesses. The management of the company has to pay attention to all these things in order to maintain harmony, productivity, high morale and teamwork, and the compensation has to be such that people are committed for the long term. By 2005, I was more interested in being a Lone Ranger and not having so many committees to deal with, including the Board of Directors, compensation committee, and all the other little bureaucratic things that have a way of creeping into companies. My style is to make decisions instantly. I found it a lot easier to do this when I was alone.

In understanding disruptive innovation, one only needs to examine the manufacturing industries in the United States. Lean manufacturing and robotics basically have increased productivity by large factors. They accomplish reduced labor by using robotics. When I observe the construction industry in this light, it appears to be the most resistant-to-change industry of all. In some ways, I am surprised that it is still in business. It is a prevailing opinion in many circles in the United States that building codes and building officials can stifle innovation. These barriers are strong in many areas, especially in large cities where entrenched special interest groups control certain systems within buildings and do not allow new and innovative products, processes and materials to gain acceptance and approval. Cities like New York, Boston, Philadelphia, and Chicago have always had entrenched rules, regulations, union jurisdictions, and work rules blocking innovations.

In the area of seismic building codes, the strong regulation environment in California, as compared to the anti-regulation environment in the Southeast, is the principal reason the nation fares so poorly in hurricanes and is highly likely to do well in earthquakes. According to my friend and colleague Chris Rojahn, Executive Director of Applied Technology Council (ATC), seismic building codes and their enforcement are clearly and emphatically the most effective way to reduce earthquake damage and losses. The founding mission of the ATC, which I have been involved with since the early 1990s, is to improve seismic building codes and their enforcement. At about the time I joined the ATC board, it was the desire of many people on the board to broaden the organization's mission to address wind, coastal flooding and riverine flooding hazards. As of today, this mission has been accomplished.

Chris vividly recalls the statement by a leading local structural engineer (forensic specialist) at an ATC symposium commemorating the 20th anniversary of Hurricane Hugo in Charleston, SC. The speaker said about the compliance of building codes in the Southeast, "Seventy percent of the buildings inspected were not compliant to wind provisions in building codes."

In my personal efforts with my new companies to create modular off-site pre-fabricated components, both modular boxes and panelized systems, the question continues to come up about whether local jurisdictions allow inspection of these components at off-site locations and eliminate the need for traditional on-site inspections. This is a perfect example how local building codes can truly stifle the ability of the American construction industry to deliver faster, more economical, and more innovative projects.

22
New Companies

Charles H. Thornton and Company

I ESTABLISHED MY FIRST COMPANY, Charles H. Thornton and Company, LLC, as a vehicle to achieve certain financial goals and personal and family financial planning. When people say that I am supposed to be retired, I paraphrase the quote of Mark Twain, "Rumors of my retirement are greatly exaggerated." In addition to planning succession in business, I decided to plan succession in my family in regard to their role in my new companies. I have a very talented family. My wife Carolyn has an MBA and is a marketing research pro. Diana, our oldest child, is a graduate of Gettysburg College and is great at business development and very aggressive in whatever she does. She became the unsalaried president of CHT and Company and Regional Coordinator for ACE Northeast. Our second child, Katherine Ann, is a graduate of Babson College with a major in accounting. Following graduation, she immediately went to work for Ernst & Whinney, which became Ernst & Young. She is now CFO of CHT & Company, LLC.

Our third child, Charles, a graduate of the University of Hartford, became very successful and satisfied as Senior Vice President in Romani

Associates, and Icon Venue, doing program and project management for sports facilities. He has been a great resource to me when it comes to program management, cost estimating, scheduling, and some of the other aspects of the construction industry, other than engineering. Our fourth child, Rebecca, was in college when I opened my new companies. She graduated from Washington College with a degree in sociology and went on to complete her master's degree in sociology with a minor in criminology at George Washington University. She is an excellent writer and has taken an interest in pro bono law and historic preservation. She manages our 17-acre spread and her two horses in Maryland.

Finally, my brother Bill, a very well respected structural steel expert, is chairman of the Manual and Textbooks Committee for the American Institute of Steel Construction. At the time I started my companies, he was chief engineer of Cives Steel Corporation and continues to be a consultant with them today.

In addition to family, Larry Hine, former CFO of Thornton Tomasetti, joined me as a financial consultant with my companies. CHT & Company offers conceptual structural engineering, structural steel consulting, management consulting, financial advice and human resources consulting.

TTG

The second company, TTG Corporation, was the disruptive innovation company. Incremental innovation doesn't take you very far – it helps to improve systems and processes, but it really doesn't catapult you into the future. Disruptive innovation, however, is the new idea of the future. In 2005, Carolyn and I had lunch with architect and inventor, David Termohlen, with whom I had worked in the early 70s. At this lunch at the Bellagio Hotel in Las Vegas, we discussed restarting his building system originally called International Environmental Dynamics (IED) System. Between 1965 and 1974, IED constructed seven projects in United States.

In 2005, we formed TTG with three investors to help. The company, which included Don Blackwelder, started redesigning condominium projects all over the United States utilizing the IED System, now called the TTG System. When we started TTG, the typical response during the first presentation to a developer, client, or contractor was, "We don't care that you've done seven projects in the United States in the past; come back when you've done the first one in the modern age." Between 2005 and 2008, major construction managers were all too busy to talk to me. Every major construction manager to whom I showed this pushed back before they even understood the benefits of the system. Introducing innovation into the archaic construction industry is like pushing a rope uphill. I would rather pull a rope. After the financial debacle of 2008, the entire American construction industry was stopped dead in its tracks. TTG had used up its money without winning a project.

The Russian Mission to the UN was constructed in the early 70s by a pretty large construction management company in New York City utilizing the International Environmental Dynamics (IED) System. Later, when I was working on projects in Nevada, the developer, who was based in New York, told me to talk to the construction management company. I set up a meeting with four top guys from this construction management company, but never told them that their company had built the Russian Mission, essentially using the TTG System. When I introduced the TTG system with all the new bells and whistles, including three-dimensional Building Information Modeling (BIM) and Computer Numerical Controlled (CNC) downloads to fabricate the steel, they told me that everything I was proposing to do would never work in New York City and that I was crazy. I then turned on my laptop and said, "I would like you to watch a video of the project that you built in New York City using the exact same system in 1974 – the Russian Mission to the UN." Their response was, "That's not fair, you didn't tell us and it didn't go as well as it was supposed to anyway." I then asked them how they would know that if they didn't even remember that they built the building.

By 2013, everything changed. Now, construction is too expensive for the average consumer or buyer and it takes too long to build. The TTG System cuts the time in half and cuts the costs by 20 to 40 percent. Construction management companies are changing their minds, barriers are breaking down, and more innovation is being accepted.

STRAAM

At the same time, I met Dr. Alan Jeary in Hawaii and invited him to come to New York to discuss his very innovative and game changing dynamic measurement systems. During the first meeting with Alan in New York, I asked him what he did. In very simplistic terms, he said I do structural risk assessment and management. Alan could take the structural heartbeat of bridges, buildings, and other structures. I wrote down STRAAM and we formed a new company and were off and running. We trademarked the term "structural cardiology" as SKG, just like an EKG for humans. Alan and his family moved to the New Jersey-New York area and we started to grow STRAAM. I am Chairman, Jeff Matros is President and CEO, and Alan is chief technology officer. STRAAM is growing very fast, but has issues with cash flow because it is in the infrastructure, which is a slow to pay industry. STRAAM is looking into alternatives of being acquired by several publicly traded and foreign companies. In the future, STRAAM technology has opportunities to revolutionize structural health monitoring of all kinds of structures, including buildings, bridges, and dams.

AECOS

In 2011, I reconnected with Brian Howells, who I had met when he was running Worldwide Real Estate for JP Morgan Bank. We formed

Automated Environmental Construction Systems (AECOS), an association of Partecnix, an automated parking system, which is controlled by Brian, and TTG, which is controlled by me.

In 2012, we were fortunate to be invited to meet with a Fortune 50 company on the West Coast that was interested in velocity, quality, and cost savings. We were successful after we redesigned one of their large industrial buildings and cut the steel weight by 52 percent, using our unique constructability approach. This approach was very successful on all of the long span arena roofs that Thornton Tomasetti did after 1989. Any engineer who designs a long span structure has to be involved in means and methods of constructability issues. Unfortunately, the AIA and American Consulting Engineers Council (ACEC), as well as insurance companies in the engineering and design industry, work based upon fear of liability, which has led to taking the engineer out of the constructability aspects of the project. This is wrong. Brian and I went to this meeting and were invited to begin work on a large office and laboratory complex in India. This was the beginning of our success.

Today, we are under contract for the project in India and we are about to commence contracts for many other similar projects in the United States, ranging from hospitals, medical office buildings, student housing projects, mid to high-rise residential rentals and condos, hotels, office buildings, and senior living facilities. As I sit here in October 2013 finishing this book, our management team with AECOS, including Brian Howells, President and CEO; Anthony Kelly, Director of Design; Steve Houston, Director of Logistics and Operations Domestically and Overseas; Graham Stewart, Vice President of MBP; Len Neuhaus, Chief Financial Officer (CFO) and Harry Spring, Managing Principal, WASA, is on its way to Bangalore, India. We finished a pricing package to obtain a guaranteed maximum price in July 2013. We are currently interviewing three very large prime contractors in India and will select a prime contractor by November 2013. Construction will start on or about January 1, 2014 and will be completed by mid-December 2014.

The project is a 621,000 square-foot, 10-story building and will be open before January 1, 2015. A conventional cast-in-place concrete structure would take at least 26 months. So the hallmarks of the AECOS/TTG System are shorter construction period, earlier occupancy, reduction in interest carry on the construction loan, reduction on the General and Special conditions, reduction in escalation costs, increased early revenue, and elimination of the rental of temporary space for all employees expecting to move into the new building. With TTG, all work is done on the ground, therefore, the quality is better, inspection is easier and less expensive, and since all workers are no more than six feet off the ground, OSHA fall protection is not required. As we move into this new era it will be proven that this is safest way to build a building.

The additional benefits with the TTG System are the central plant chillers and cooling towers and mechanical rooms which are all prefabricated off-site and placed on the roof at ground level. By the time we lift the roof and top three floors, the air-conditioning system and/or heating system is activated depending upon the climate. Since the system has no columns, the efficiency factor is 3 percent, which means that we can build 3 percent less gross square footage than a conventional reinforced concrete building. Furthermore, in an office building, from the seat-count point of view, there are 7 percent more seats in the same area than with conventional cast in place construction. Combining seat count and the absence of columns produces a 10 percent improvement in efficiency over the conventional concrete building. To build a comparable building in a conventional concrete system would require 11 stories or 62,100 more square feet.

When construction of the project in Bangalore commences and we have Webcams, animations, and videos of the rapid construction, AECOS will take off like a rocket. Persistence, passion, and flexibility are what it takes to do and succeed at disruptive innovation.

23
Growing ACE Mentor Program

THE NEXT THING I DID was concentrate on the growth of the ACE Mentor Program. In 2005, Norbert Young, president of McGraw-Hill Construction, and I approached the Associated General Contractors (AGC) to solicit their sponsorship of ACE. We flew to the AGC meeting in Las Vegas in March 2005 and were successful in getting AGC to become a national sponsor of the ACE Mentor Program of America. At this point, ACE was only twelve-years-old. Turner Construction Company, Gilbane Construction Company, McGraw-Hill, and Emcor were already major sponsors. AGC's board voted to become an additional national sponsor.

Today, a total of 40,000 inner-city high school students in 106 cities around the nation have been introduced to the challenges and rewards found in careers in the industries of architecture, construction, and engineering. ACE has awarded more than $14 million in scholarships – all generated by local affiliate breakfasts, lunches, and dinners. The ACE National Board of Directors led by me and Tom Gilbane, Jr., Chairman and CEO of Gilbane Building Company, as co-chairmen; Peter Davoren, CEO of Turner Construction Company as vice chairman; Murray Savage,

CEO of Professional Services Industries, Inc., Treasurer; John Strock, Executive Director, ACE Mentor Program of America, Inc.; Richard Allen, Senior Vice President and COO of Stantec; Charlie Bacon, CEO of Limbach Facility Services; Joan Calambokidis, President of International Masonry Institute; Mark Casso, President of The Construction Industry Round Table; Patricia Coleman, Principal of Thornton Tomasetti; Sandy Diehl, CEO of SD Global Advisors; Tony Guzzi, CEO of Emcor Group; Hank Harris, President and Managing Director of FMI Corporation; and Ross Myers, CEO of American Infrastructure, and are committed to taking this program to many more students per year. The ACE Mentor Program is growing rapidly and we are well on our way to accomplishing this goal.

There are a number of exciting opportunities for the future of the ACE Mentor Program. The overall game plan is to add more regional managers for the Program, starting in the central region of the United States, probably in Houston, Texas. The American Concrete Institute, which is active in Central America, Mexico, and South America, is actively pursuing growing ACE into Central America, South America, and Mexico. As time goes by, I expect ACE will be present in the United Kingdom, Saudi Arabia, India and many other places.

In 2012, President Barack Obama awarded the ACE Mentor Program the Presidential Award for Excellence in Science, Mathematics, and Engineering Mentoring – the nation's highest award for mentoring in the science, technology, engineering and mathematics (STEM) fields. In my acceptance speech for the award at the White House, I emphasized that the ACE Mentor Program is both scalable and sustainable, funded almost entirely by the industry we represent. The ACE National Program performs surveys of recent ACE graduates. The results are astounding – 98 percent of high school seniors who completed ACE in 2013 graduated from high school as compared to the national average of 78 percent in 2010. Furthermore, 95 percent of high school seniors who completed ACE in 2013 enrolled in college immediately after graduation, compared to the national average of 66 percent in 2012. More specifically, 69 percent of high school seniors who completed ACE in 2013 enrolled in college with a declared major in architecture, construction or engineering. The most recent surveys performed by the ACE National Mentor Program are listed in the appendices of this book.

24
Risk and Perceived Risk Vs. Opportunity

PREVIOUSLY, I DISCUSSED STARTING CHT and Company, TTG Corporation, STRAAM Corporation, and AECOS. When I started these companies, very few people believed that the end goal could be achieved. These people are the doubting Thomases, the skeptics and the Old Order, as described by Machiavelli in the quote opening this section of the book. As of the fall of 2013, all of these start-up companies continue to be successful.

I believe that most risk is perceived. If anyone has the guts to manage these risks – they cease to become risks, but instead, become opportunities. The following is a list of the risks I have experienced throughout my life:

Risk # 1 – Childhood in Clason Point, Bronx, NY

All the childhood activities in which we participated had a certain degree of risk. With the encouragement and training of my father in boating activities, hiking, cycling and by using common sense, I don't believe there was ever any risk. We took safe boating courses and learned how to handle slow boats and very fast boats. As far as I'm concerned, we really controlled and managed all of those risks.

I feel that overprotective parents produce just the opposite result. It's good to allow children to experiment with multiple activities and enable them to experience interesting situations and how to figure out how to get out of them after getting into them.

Risk # 2 – Education

The progression from PS 107 to PS 69 and back to PS 107, then onto Iona Prep, Manhattan College and NYU was a risk. Maybe it was risk of failure or fear of failure. Each time I moved up the stepping-stones of education, the competition got fiercer. Bill and I were two of the few who went to college from our neighborhood. Think about the fact that at Iona Prep, most of the students came out of the top of their classes in their private schools. At Manhattan College, most of the students came out of the top of their classes in their private schools. In order to succeed in college, I needed to exhibit a combination of hard work, strong study habits and the ability to size up the faculty, communicate with the faculty by showing that I was both interested and interesting. I always talked to professors after the class and always ended up with a slightly higher grade. Having taught at Cooper Union, Pratt, Manhattan College, Princeton, and Catholic University, I always rewarded the students that really showed an interest and they received a slightly higher grade. The people that hid or sat up front looking straight down hoping I wouldn't notice them generally had a harder time getting decent grades.

Embarking on a PhD program at NYU was a daunting experience. A PhD thesis, which has to be a significant contribution to the profession, also has to be original. I think if the average PhD student understood the long dark tunnel that it takes to get through the process, he or she might stop. Although the student doesn't know it, at his final thesis defense, he actually knows more about that subject than anybody in the room. His real fear is that some rogue jerk professor, who's got it in for somebody, maybe even him, will show up and challenge what he has undertaken for his thesis. I found that everyone who was congenial, friendly, and took the right courses with the right committee members, all received their PhDs.

Risk # 3 – Marriage

With the high divorce rate in the United States fear of failure leading to a divorce, or losing a child, or having a child grow up with no sense of civic duty or obligation could be a big problem. As we learn in parenthood, no one teaches us how to be a parent. I found that using common sense and listening to the wise and sage advice from parents and grandparents and in-laws, really worked.

Risk # 4 – Starting an Engineering Career

I benefited greatly from joining the LZA firm in its infancy, at about the point when LZA won the design commission for 14 pavilions at the 1964 to 1965 World's Fair in New York City. This was a unique opportunity as each of these 14 pavilions had unique structural solutions with plenty of risk. However, my education, receiving my master's degree while working on the spectacular projects from 1961 through 64, especially prepared me for what Lev Zetlin Associates, Thornton Tomasetti and I accomplished in my career.

During this period, I was taking all the great mechanics courses at New York University's various departments including civil/structural, aeronautical, mechanical, and mathematics in the mathematics department and at the Courant Institute. I had the perfect theoretical training, technical back up, and practical experiences to solve any technical problem perceived or real. I also developed the ability to locate and hire the right people. The team I selected in 1968, at age 28, for the American Airlines Super Bay Hangar would have been capable of doing almost any project. Selection of the right team with the right education and the right communication skills is the key to success in probably all areas especially engineering.

Risk # 5 – Remarriage

Proposing marriage to Carolyn, my second wife, who had never been married before, while living with three children, a mother and a father and mother-in-law in the same house, appeared to be a risk. Carolyn and I identified the risks to a blended family, so we took steps to eliminate them. This involved allowing the children to stay in the family home with the grandparents to finish their schooling and Carolyn and I having the chance to begin our marriage together, essentially alone, with the children visiting us regularly. The whole situation worked out with great benefit for everyone, including the grandparents, whom we called "the senior citizens," the children, Carolyn, and myself.

Risk # 6 – Buying LZA

Buying LZA from Gable Industries with Richard L. Tomasetti, a person of great trust and integrity, was in my opinion not a risk. The thought of owing $500,000 in 1977 was a little daunting. Richard and I had the ability to split the ownership 50-50, hire the right people, and manage the team with spirit and trust and with great rewards for those who performed and dispatching those who didn't. This was a leveraged buyout. We had five years to pay off $500,000; it required no money down and no personal guarantees. We paid it off in 2 ½ years.

Risk # 7 – Selling Thornton Tomasetti Internally

As described in this book, Richard and I decided to transition the company over a 10-year period from 1995 to 2005, by selling a majority of the shares to seven people in the first five years and the balance in the second five years. These seven people were the people who truly helped Richard and me get the company in the position we were – an extremely well-known and respected structural engineering firm. Allowing them to have a majority of the stock would be considered a risk to most people, but it was never a risk to me because these people were hand-picked, they were very talented, they were very honest and ethical people and I never felt the was any risk at all. They took the controls and took off – because we empowered them. As I moved to Washington in 2000, I encouraged them all to start running the company, which they did with perfection.

As you have read this book it's hard to avoid the conclusion that maintaining relationships from early in your career to the end of your career, leads to success. Maintaining relationships is the key to successful sustained growth in a business because as these people grow and move into very important positions in life they take you along with them. As the company grew and expanded, I made it a point to stay in touch with everybody. If you take a person like George Pavarini, whom I first met in 1963 and still talk to several times a year, you have a great example. I still stay in regular contact with Ken Hiller, whom I met in 1968 at the beginning of the American Airlines Super Bay Hangar Project. The same is true of George Feddish, Frank Marino, Joe Denny, Joe Thelen, Robert Selsam, and Ed Reidel, to name a few.

The other hallmark of being successful is to be fun to work with. Structural engineering is a serious business and we take some awesome responsibilities. But the engineers and the engineering firms that are fun to work with and fun to play with are the ones who succeed. People make fun of engineers as being absent of personality and creativity. This is not true, but sometimes this is the impression. Building codes and standards and engineering methodologies tend to really dampen creativity and enthusiasm in engineering. Building codes are minimum standards. They are just guidelines. If you first stick to the principles of physics, science and engineering, building codes and code officials should not be a hindrance to progress. If you think about the overused out-of-the-box term, it is really not out-of-the-box. It's just disruptive innovation. It is being fresher than everyone else. It is trying new ideas.

I hope you have enjoyed this book and look forward to discussing it with you at your convenience. Websites showing information about

Thornton Tomasetti and the new Charles H. Thornton companies are listed in the appendix. I urge you to contact me either by phone or by e-mail and let's have a spirited discussion about how we can make the world better and engineering more exciting for young people and practicing professionals.

Epilogue

I HAVE ALWAYS APPRECIATED ALFRED Lord Tennyson's poem, "Crossing the Bar." I have crossed many bars sailing and in life – CHT

Crossing the Bar

by Alfred Lord Tennyson

Sunset and evening star,
And one clear call for me!
And may there be no moaning of the bar,
When I put out to sea,

But such a tide as moving seems asleep,
Too full for sound and foam,
When that which drew from out the boundless deep
Turns again home.

Twilight and evening bell,
And after that the dark!
And may there be no sadness of farewell,
When I embark;

For tho' from out our bourne of Time and Place
The flood may bear me far,
I hope to see my Pilot face to face
When I have crost the bar.

- Alfred Lord Tennyson

Appendix – Awards & Accomplishments

Awards

Member of The Moles, since 1987

James F. Lincoln Arc Welding Foundation Gold Award, 1986 & 1988

American Society of Civil Engineers, Met Section, Civil Engineer of the Year, 1990

The Concrete Industry Board Leader of Industry Award, 1991

Member of the National Academy of Engineering, since 1997

Member of National Academy of Construction, 1997

Honorary Member of the American Society of Civil Engineers, 1999

Honorary member of American Institute of Architects, 2006

Honorary member of the International Union of Bricklayers and Allied Craftworkers

National Institute of Building Sciences Honor Award, 2002

Leonardo de Vinci Award for Leadership and Management Excellence by the Professional Services Management Association

Benjamin Franklin Medal in Civil Engineering by The Franklin Institute, 2003

Golden Eagle Award by the Society of American Military Engineers

American Society of Civil Engineers Hoover Medal for Distinguished Public Service, 2002

Engineering News-Record Top 25 News Makers, 1978, 1996 and 2000

Engineering News-Record Top 125 People of the Past 125 Years, 1999

Engineering News-Record Award of Excellence, 2001

Michelangelo Award by The Construction Specifications Institute, 2006

National Building Museum Henry C. Turner Prize for Innovation in Construction Technology, 2008

Construction Industry Institute (CII) Carroll H. Dunn Award of Excellence, 2009

American Institute of Steel Construction (AISC), Designer Lifetime Achievement Award, 2010

Faziur R. Khan Lifetime Achievement Medal from the Council on Tall Buildings and Urban Habitat (CTBUH), 2012

Presidential Award for Excellence in Science, Mathematics, and Engineering Mentoring, 2012

American Society of Civil Engineers Outstanding Projects And Leaders (OPAL) Award, 2013

Honorary Doctorate Degrees

University of Connecticut, 2001

Clarkson University, 2004

Rensselaer Polytechnic Institute, 2004

Manhattan College, 2013

Appendix – Websites

Charles H. Thornton & Company	chtandcompany.com
STRAAM Corporation	straamllc.com
AECOS Companies	aecosltd.com
ACE Mentor Program of America	acementor.org
Charles H. Thornton Artwork	charleshthorntongallery.com
Charles H. Thornton Book	LifeOfElegantSolutions.com
	ALifeOfElegantSolutions.com
	CharlesHThornton.com
Thornton Tomasetti	thorntontomasetti.com

Photographs

Evelyn Thornton with Son Charlie

Charles Thornton, Sr. with Sons
(Left to Right) Charlie, Robert, and William

Phi Kappa Theta Fraternity Civil Engineering Classmates from Manhattan College in 1961. 1st Row: 4th from Left, Mike Bellanca; 6th from Left, Jack Mcnamara. 2nd Row: Far Left, Charlie Thornton; 4th from Left, Gus Fruh. 3rd Row: Far Left, Richard Tomasetti

Manhattan College Senior Portrait, 1961

Charles Thornton, Sr. with Granddaughter Diana, 1962

Kathy, Charlie III, and Diana, 1968

Patricia with Children (Left to Right) Charlie III, Kathy, and Diana at Storm King, West Point, NY, 1970

Charlie Thornton's First Sailing Cruise, Shelter Island, June 1979

Carolyn and Charlie Thornton with (Left to Right) Kathy, Diana, and Charlie III in Essex, Ct on Their Wedding Day, 1981

Carolyn and Charlie Thornton Sailing in Antigua on Raffles Light

Male Bonding in the Caribbean. (Left to Right): Michael Thornton, Joe Denny, Charlie Thornton, William Thornton, Ed Reidel, and Harry Armen

Charlie Thornton and Richard Tomasetti at the Brennan Beer Gorman Regatta

Charlie Thornton with Daughter Rebecca "Becky" in Jiminy Peak, 1986

Charlie Thornton's Artwork

Becky Showing her Horse, Nantucket

Becky's Graduation from Gunston Day School (Left to Right) are Diana, Becky and Evelyn Thornton, 2004

The Thornton Family (Left to Right) Back Row: Brian O'Connell, Brandon O'Connell, Kaitlin O'Connell, Casey Eidenshink, Diana Thornton Eidenshink, Ryan Eidenshink. Middle Row: Carolyn Thornton, Charlie, Becky Thornton, Andrew Eidenshink, Charlie Thornton III. Front Row: Kathy Thornton O'Connell, Charlie Thornton IV, Meghan Thornton, and Kate McHugh Thornton.

Charlie Thornton in 2004 with ACE Mentor Program Students in Easton, MD (Left to Right) Cory Schwarm, Sts. Peter and Paul High School; Charlie; Carolina Santos, Project Manager at Willow Construction, LLC; and Leanna Isom, Easton High School (Photo Courtesy of Willow Construction, LLC)

Index

C

D

E

M

U

V

Charles H.Thornton (Photo Courtesy of Amy Blades Steward)

Charles H. Thornton, PhD, PE, was a founding principal and formerly chairman of the structural engineering company Thornton Tomasetti. He has been an adjunct professor at The Cooper Union, Pratt Institute, Manhattan College, Princeton University, and Catholic University. Dr. Thornton founded the ACE Mentor program, a nationwide non-profit organization offering guidance and training in architecture, construction and engineering to more than 40,000 inner city high school students in 106 cities across the United States. He lives in Maryland with his wife.

Amy Blades Steward (Photo Courtesy of Melissa Grimes Guy Photography)

Amy Blades Steward is founder of Steward Writing and Communications, a public relations firm in Easton, MD. Her company focuses on copy writing and editing services for non-profit and for profit companies, small businesses, and local governments. She has written non-fiction articles for national, regional, and local publications for over 30 years. Although a lifelong storyteller, this is Steward's first book. She lives with her family in Maryland.

Made in the USA
Las Vegas, NV
12 January 2024

84264919R00156